London Mathematical Society Student Texts 15

Presentations of Groups

D. L. JOHNSON
Department of Mathematics, University of Nottingham

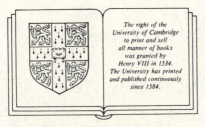

*The right of the
University of Cambridge
to print and sell
all manner of books
was granted by
Henry VIII in 1534.
The University has printed
and published continuously
since 1584.*

CAMBRIDGE UNIVERSITY PRESS
Cambridge
New York   Port Chester
Melbourne   Sydney

Published by the Press Syndicate of the University of Cambridge
The Pitt Building, Trumpington Street, Cambridge CB2 1RP
40 West 20th Street, New York, NY 10011, USA
10 Stamford Road, Oakleigh, Melbourne 3166, Australia

First published 1990

Printed in Great Britain at the University Press, Cambridge

Library of Congress cataloguing in publication data: available

British Library cataloguing in publication data available

ISBN   0 521 37203 8   hard covers
ISBN   0 521 37824 9   paperback

This book is dedicated to the
memory of

ROGER LYNDON

who inspired, befriended and
encouraged us all, and who said,
''Perhaps combinatorial group theory
  is just a state of mind.''

CONTENTS

PREFACE

This book has evolved from a course of lecturers for final-year undergraduates and first-year postgraduates. This was given at the University of Nottingham, first in the session 1972-73 and many times since, and at Busan National University and the University of Adelaide. The material is accessible to any one with the mathematical maturity consistent with first courses in linear algebra, group theory and ring theory, and is primarily intended as an introduction to the subject known as combinatorial group theory. This abuts on other branches of group theory: finite, infinite, homological, and computational. A secondary aim is to introduce a wide variety of examples of groups and types of group.

No attempt at completeness is feasible in a work of this size and scope. Major omissions, or taking-off points, include small cancellation theory, decision problems and embedding theorems as well as such other topics of historical and current importance as commutator calculus, Fuchsian groups, braid and knot groups, one-relator groups, free products with amalgamation, HNN-extensions, and geometric methods.

Compared to earlier editions (nos. 22 and 42 of the LMS Lecture Notes Series) the "core material" in Chapters 1, 2, 4 and 6 has been preserved more or less intact, as have the contents of Chapters 7, 8, 10 and 15. Chapters 5, 9 and 13 have been substantially revised, and Chapters 3, 11, 12 and 14 are completely new.

My thanks are due to a host of people (not least the students who have attended the course) whose names are to be found scattered through the pages of the text. It is a pleasure to acknowledge a special debt of gratitude to Sandy Green for first introducing me to research mathematics, and to Roger Lyndon, Bernhard Neumann, Jim Wiegold, Mike Newman and Edmund Robertson for their support and encouragement over the years. Thanks also to Mary Mattaliano and Krystyna Glowzgewska for making such a good job of the typescript and to David Tranah and all at C.U.P. for their efficient handling of the production.

CHAPTER 1

FREE GROUPS

The fundamental notion underlying the theory of group presentations is that of a free group. Roughly speaking, a group F is called free if it has a subset X with the property that every element of F can be written uniquely is a product of elements of X and their inverses. The uniqueness here means that if two such products look different, then they are different, so that no non-trivial relations hold between elements of F. This is made precise in the definition given below, which suggests (correctly) that the idea of freeness is also applicable in algebraic situations other than group theory.

1.   Definition and elementary properties

Definition 1. A group F is said to be *free on a subset* $X \subseteq F$ if, given any group G and any map $\theta : X \to G$, there is a unique homomorphism $\theta' : F \to G$ extending θ, that is, having the property that $x\,\theta' = x\,\theta$ for all $x \in X$, or that the diagram

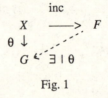

Fig. 1

is commutative. Then X is called a *basis* of F and $|X|$ the *rank* of F, written $r\,(F)$.

Remarks. 1. This accords with our intuitive idea of freeness in so far as:

(i)   if there were a "relation" $w = w'$ among the members of $X^{\pm} = \{x,\, x^{-1} \mid x \in X\}$, then we could find a group G with corresponding elements (under θ) for which that relation does not hold; then θ' would have to map $e = w'\,w^{-1} \in F$ to an element of G other than the identity.

(ii) if the elements of X did not generate F, the extension of θ would only be defined as far as the subgroup $<X>$ of F generated by X and thereafter would be arbitrary, violating the uniqueness.

In this sense, then,

$$\text{existence of } \theta' => \text{ no relations in } X^{\pm},$$
$$\text{uniqueness of } \theta' => X \text{ generates } F.$$

2. There is a strong analogy here with the notion of "extension by linearity" in the theory of vector spaces: if B is a basis for a vector space V and $\theta : B \to W$ any map into a vector space W, there is a unique extension of θ to a linear transformation from V into W. Here, the existence and uniqueness of the extension are guaranteed by the two defining properties of a basis, namely, that the elements of B are linearly independent and span V, respectively.

3. Prefacing the word "group" by the adjective "abelian" in the two places where it occurs in the above definition yields another, that of *free abelian group*. This is even more like the situation in linear algebra, for if A is a free abelian group with basis $X = \{x_1, ..., x_n\}$ say, then every element of A is uniquely expressible in the form $\sum_{i=1}^{n} k_i x_i, k_i \in \mathbb{Z}, 1 \leq i \leq n$. Such an expression is called a $\mathbb{Z}$-linear combination, and we have an isomorphism

$$\left. \begin{array}{rcl} A & \longrightarrow & \mathbb{Z} \oplus \mathbb{Z} \oplus ... \oplus \mathbb{Z} \\ \Sigma k_i x_i & \longmapsto & (k_1, k_2, ..., k_n) . \end{array} \right\}$$

In fact, the only conceptual difference between a free abelian group of rank n and a vector space of dimension n is that in the former, the coefficients $\mathbb{Z}$ do not form a field.

4. Now let F be an arbitrary group, $X \subseteq F$, and G any group. Let Hom (F, G) denote the set of homomorphisms from F into G, Map (X,G) the set of maps from X into G, and

$$\left. \begin{array}{l} \rho : \text{ Hom } (F,G) \to \text{Map } (X,G) \\ (F \overset{\phi}{\longrightarrow} G) \mapsto (X \overset{inc}{\longrightarrow} F \overset{\phi}{\longrightarrow} G) \end{array} \right\} \tag{1}$$

the restriction map. Then

$$\rho \text{ is surjective} \Leftrightarrow \forall \theta, \exists \theta' \text{ as in Definition 1,}$$
$$\rho \text{ is injective} \Leftrightarrow \theta', \text{ if it exists, is unique.}$$

Thus, F is free on X if and only if the map ρ of (1) is a bijection for any group G.

5. While a free group may have many different bases, they all turn out to have the same number of elements, so that the rank is well-defined. This will be proved directly, along with the converse that a free group is determined up to isomorphism by its rank. The

existence of free groups is then proved by explicit construction.

Lemma 1. If F is free on X, then X generates F.

Proof. Let $H = \langle X \rangle := \cap \{K \leq F \mid K \supseteq X\}$, and let $\theta : X \to H$ denote inclusion, with $\theta' : F \to H$ the corresponding extension. Letting $\iota : H \to F$ denote inclusion, we see from the picture that $\theta' \iota$ extends $\theta \iota = \text{inc}$. But so does the identity map 1_F. By uniqueness, $\theta' \iota = 1_F$, whence $F = \text{Im } 1_F = \text{Im } \theta' \iota = \text{Im } \theta' \subseteq H$.

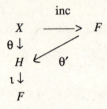

Fig. 2

Proposition 1. If F_i is free on X_i $(i = 1,2)$ and $F_1 \cong F_2$, then $|X_1| = |X_2|$.

Proof. Apply Remark 4 above with $G = Z_2$ the group of order 2. Since $F_1 \cong F_2$, $|\text{Hom } (F_1, Z_2)| = |\text{Hom } (F_2, Z_2)|$, whence $|\text{Map } (X_1, Z_2)| = \text{Map } (X_2, Z_2)|$. But for any sets B, C of cardinalities b, c, respectively, $|\text{Map } (B,C)| = c^b$. So in this case, $2^{|X_1|} = 2^{|X_2|}$, and the result follows by taking logs to the base 2.

Proposition 2. If F_i is free on X_i $(1 = 1,2)$ and $|X_1| = |X_2|$, then $F_1 \cong F_2$.

Proof. Assume $|X_1| = |X_2|$, so that there is a bijection $\kappa : X_1 \to X_2$. Let α, β be the extensions given by:

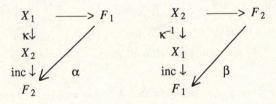

Fig. 3

Now for all $x_1 \in X_1$, $x_1 \alpha\beta = (x_1\kappa)\beta = (x_1\kappa)\kappa^{-1} = x_1$, so that $\alpha\beta : F_1 \to F_1$ extends inc $: X_1 \to F_1$. But so does the identity map 1_{F_1} on F_1, whence $\alpha\beta = 1_{F_1}$ by uniqueness. Similarly, $\beta\alpha = 1_{F_2}$, and so α is an isomorphism.

2.   Existence of $F(X)$

Let X be any abstract set. There follows a recipe for constructing a free group $F(X)$ containing X as a basis.

Step 1. First form another copy of X, $\hat{X} = \{\hat{x} \mid x \in X\}$ (whose elements will later become the inverses of the elements of X), and consider their union $X^{\pm} := X \cup \hat{X}$.

Next form the *words* $W_n = (X^{\pm})^{\times n}$ of length $n \geq 0$ in $X^{\pm}$, which are just n-tuples of elements of $X^{\pm}$. Thus,

W_0 consists only of ( ), the *empty* word, usually written e,

W_1 consists of (x), $(\hat{x})$, $x \in X$, and so looks like $X^{\pm}$,

W_2 consists of pairs (x,y), $x,y \in X^{\pm}$, and so on.

Now discard all words containing a pair $x, \hat{x}$ (for the *same* $x \in X$) in adjacent positions, in either order. The remaining words, namely those without this property, are called *reduced*: let $\tilde{W}_n$ denote the set of reduced words of length n.

Finally, define $F(X) = \underset{n \geq 0}{\cup} \tilde{W}_n$.

Step 2. For $F(X)$ to be a group, we need a binary operation on $F(X)$, defined roughly as "juxtaposition plus cancellation", and precisely as follows: given

$$a = (x_1, ..., x_l) \in \tilde{W}_l, \ b = (y_1, ..., y_m) \in \tilde{W}_m , \tag{2}$$

put

$$ab = (x_1, ..., x_{l-r}, y_{r+1}, ..., y_m) , \tag{3}$$

where r is the largest value of k for which none of (x_l, y_1), (x_{l-1}, y_2), ..., (x_{l-k+1}, y_k) is reduced. This condition guarantees that $ab \in \tilde{W}_{l+m-2r}$. Of course $r \leq \min(l,m)$ and equality may occur.

Closure being assured, it remains to check the other axioms for a group. Clearly, e is the identity, and

$$(x_1, ..., x_l)^{-1} = (\hat{x}_l, ..., \hat{x}_1) ,$$

interpreting $\hat{x} = x$ when $x \in X$. For the associative law, let a, b, ab be as in (2) and (3), and let

$$c = (z_1, ..., z_n) \in \tilde{W}_n, \quad bc = (y_1, ..., y_{m-s}, z_{s+1}, ..., z_n) \in \tilde{W}_{m+n-2s} .$$

Thus we have to show that

$$(ab)c = a(bc) . \tag{4}$$

If any of a, b or c is e, this is obvious, and so we can assume that l, m, $n \geq 1$. There are three cases to consider.

Case 1: $r + s < m$. Here, both sides of (4) are equal to

$$(x_1, ..., x_{l-r}, y_{r+1}, ..., y_{m-s}, z_{s+1}, ..., z_n) \in \tilde{W}_{l+m+n-2r-2s} .$$

Case 2: $r + s = m$. In this case, both sides of (4) are equal to the product

$$(x_1, ..., x_{l-r})(z_{s+1}, ..., z_n) .$$

Case 3: $r + s > m$. Put

$$\beta = (y_1, ..., y_{m-s}), \quad \gamma = (y_{m-s+1}, ..., y_r), \quad \delta = (y_{r+1}, ..., y_m) ,$$

so that γ has length ≥ 1 by case hypothesis, and

$b = \beta \gamma \delta$, which is unambiguous by Case 1, as are

$a = \alpha \gamma^{-1} \beta^{-1}$, where $\alpha = (x_1, ..., x_{l-r})$, and

$c = \delta^{-1} \gamma^{-1} \varepsilon$, where $\varepsilon = (z_{s+1}, ..., z_n)$.

Then

$$(ab)c = (\alpha \delta)(\delta^{-1}\gamma^{-1}\varepsilon) = \alpha(\gamma^{-1}\varepsilon) ,$$

and

$$a(bc) = (\alpha \gamma^{-1}\beta^{-1})(\beta \varepsilon) = (\alpha \gamma^{-1})\varepsilon .$$

Since α and γ^{-1} are adjacent in the reduced word a, there is no cancellation in forming their product, and similarly with γ^{-1} and ε (in c), whence $\alpha(\gamma^{-1}\varepsilon) = (\alpha\gamma^{-1})\varepsilon$ by Case 1,

showing that (4) holds in every case.

Step 3. It is now appropriate to use the bijection

$$W_1 = \tilde{W}_1 \;\rightarrow\; X^{\pm} \Big\rangle$$
$$(x) \;\mapsto\; x \Big\rangle$$

to simplify notation by removing brackets and commas. Then $\hat{x}$ naturally becomes identified with x^{-1}, and in view of such rules as $x^{-1}xy = y$ and $(xy)^{-1} = y^{-1}x^{-1}$, the above definitions of product and inverse are in accordance with standard notational conventions. Reduced words are those in which no cancellation in the usual sense is possible. We retain the notion of length, and write $l(w) = n$ if $w \in \tilde{W}_n$. Note that $X \subseteq F(X)$, and that $<X>$ contains both $X^{-1} = \{x^{-1} \,|\, x \in X\}$ (by inversion) and thus every $\tilde{W}_n$ (by closure), whence X generates $F(X)$.

Step 4. Finally, it must be shown that $F(X)$ is free on X. For a given G, and a given $\theta : X \rightarrow G$, define

$$e\theta' = e, \;\; x\theta' = x\theta \;\; \text{and} \;\; (x^{-1})\theta' = (x\,\theta)^{-1} \;\; \text{for } x \in X \; ,$$

and

$$(x_1 \ldots x_l)\,\theta' = (x_1\theta') \ldots (x_l\theta'), \;\; \text{for } x_1 \ldots x_l \in \tilde{W}_l \; .$$

It is clear that θ' extends θ and that there can be at most one homomorphism with this property, as X generates $F(X)$. It remains only to prove that $(ab)\theta' = (a\theta')(b\theta')$, with a, b, ab as in (2) and (3), above. By the definition of ab, no (x_{l-i+1}, y_i) is reduced for $1 \leq i \leq r$, and so $y_i = x_{l-i+1}^{-1}$, and $y_i\theta' = (x_{l-i+1}\theta')^{-1}$ for all such i, by the definition of θ' on words of length 1. It follows from the definition of θ' on longer words that

$$(a\theta')^{-1}(ab)\theta'(b\theta')^{-1} = (x_l\theta')^{-1} \ldots (x_{l-r+1}\theta')^{-1} \, (y_r\theta')^{-1} \ldots (y_1\theta')^{-1}$$

$$= (y_1\theta') \ldots (y_r\theta') \, (y_r\theta')^{-1} \ldots (y_1\theta')^{-1}$$

$$= e \; ,$$

as required.

Theorem 1. The group $F(X)$ of reduced words in $X^{\pm}$ is free on X.

This result has two important consequences which provide an internal characterisation of free groups and a starting point for the theory of presentations, respectively.

Proposition 3. A group F is free on a subset X if and only if:

(i) X generates F, and

(ii)   no reduced word in $X^{\pm}$ of positive length is equal to e.

Proof. Let $\theta' : F(X) \to F$ be the homomorphism extending the inclusion $\theta : X \to F$. Then (i) and (ii) are respectively equivalent to the assertions that θ' is surjective and injective. On the other hand, if F is free on X, the extension $\phi' : F \to F(X)$ of the inclusion $\phi : X \to F(X)$ is clearly an inverse of θ' (using Lemma 1), while if θ' has an inverse, it must be an isomorphism, and freeness is an isomorphism invariant (see Exercise 4). Thus,

$$F \text{ is free on } X \Leftrightarrow \theta' \text{ is a bijection} \Leftrightarrow \text{(i) and (ii) hold.}$$

Proposition 4. Every group is isomorphic to a factor group of some free group.

Proof. Given a group G, let X be a set of generators for G (which always exists: take $X = G$, for example). Then let $\theta' : F(X) \to G$ be the extension of the inclusion $\theta : X \to G$. Now Im $\theta' = G$, since $<X> = G$, so that if $K = \text{Ker } \theta'$, then

$$G = \text{Im } \theta' \cong F(X) / K \ ,$$

by an Isomorphism Theorem.

3.   Further properties of *F(X)*

Several important properties of free groups depend only on elementary arguments involving the lengths of words. The examples which follow are fairly typical and are included for three reasons. a) They form a rather pleasing array of interrelated results; b) their culmination (Theorem 2) is, in several works left as an exercise; c) Theorem 2 provides a useful illustration of the power of the Nielsen-Schreier theorem (proved in Chapter 2), of which it is an easy consequence (see Exercise 2).

Let $F = F(X)$ denote the free group on a fixed set X. The following definition will be useful.

Definition 2. A reduced word $a = x_1 x_2 \dots x_l$, $x_i \in X^{\pm}$, $1 \le i \le l$, is called *cyclically reduced* if $x_l \ne x_1^{-1}$.

Now let $a = x_1 \dots x_l$ be any reduced word in $X^{\pm}$, and let $a^2 = x_1 \dots x_{l-r} x_{r+l} \dots x_l$, reduced, so that $l(a^2) = l(a) - 2r$. How big can r be? Clearly, $r = 0$ if and only if a is cyclically reduced. To answer this in general, first let $l = 2k + 1$ be odd. Then it is clear that $r \le k$, for otherwise, $x_{k+1} = x_{k+1}^{-1}$, that is, $x_{k+1}^2 = e$, which is impossible by condition (ii) in Proposition 3. On the other hand, when $l = 2k$ is even, we must have $r < k$, for otherwise $x_k = x_{k+1}^{-1}$, contrary to the fact that a is reduced. It follows that $r < l/2$, whence $a = u^{-1} \check{a} u$, where

$$u^{-1} = x_1 \dots x_r = x_l^{-1} \dots x_{l-r+1}^{-1}, \quad \check{a} = x_{r+1} \dots x_{l-r}, \tag{5}$$

with $\check{a} \ne e$ and $x_{r+1} \ne x_{l-r}^{-1}$, so that $\check{a}$ is cyclically reduced. Note that

$$l(a^2) = 2l - 2r > 2l - l = l = l(a) .$$

More generally, for $n \in \mathbb{N}$, $a^n = u^{-1} \check{a}^n u$, and since it is clear that $\check{a}^n$ is cyclically reduced, it follows that

$$l(a^n) = nl(\check{a}) + 2r > (n-1) l(\check{a}) + 2r = l(a^{n-1}) . \tag{6}$$

Thus, no elements of $F(X)$ other than e can have finite order. Groups with this property are called *torsion-free*.

Proposition 5. $F(X)$ is torsion-free.

The next point is that free groups are as non-commutative as they can possibly be, in the following sense. In any group, if two elements are powers of a common element, then they must commute. The key lemma asserts that in $F(X)$, the converse holds.

Lemma 2. Let $a, b \in F(X)$ satisfy $ab = ba$. Then there is a $c \in F(X)$ such that $a = c^k$, $b = c^h$, $h, k \in \mathbb{Z}$.

Proof. Proceed by induction on $l(a) + l(b)$. Since the result is clear when either a or b is e, the basis for the induction is established, and we may assume that $a \ne e \ne b$. Letting $a = x_1 \dots x_l$, $b = y_1 \dots y_m$, assume that $l = l(a) \le l(b) = m$, by symmetry. Now consider the equation $ab = ba$ in reduced form:

$$x_1 \dots x_{l-r} y_{r+1} \dots y_m = y_1 \dots y_{m-r} x_{r+1} \dots x_l , \tag{7}$$

where $0 \le r \le \min (l,m) = l$, by assumption. Distinguish three cases.

Case (i): $r = 0$. Then it follows from (7) that $x_i = y_i$, $1 \le i \le l$, by comparing initial segments. Then, $b = au$, with $l(u) = m - l < m$, so that $l(a) + l(u) < l(a) + l(b)$. Then $au = b = a^{-1}ba = a^{-1}aua = ua$, and the inductive hypothesis yields that a and u are both powers of some $c \in F$, whence so is $b = au$.

Case (ii): $r = l$. Here, $y_i = x_{l-i+1}^{-1}$, $1 \le i \le l$, and $b = a^{-1}u$, with $l(u) = m - l < m$. It follows as in case (i) that a^{-1} and u commute and so are again powers of a common c, and $b = a^{-1}u$ is too.

Case (iii): $0 < r < l$. In this case,

$$x_1 = y_1, \quad x_l = y_m, \quad x_l = y_1^{-1}, \quad y_m = x_1^{-1} \ ,$$

whence

$$a = x_1 a' x_1^{-1}, \quad b = x_1 a' x_1^{-1} \ ,$$

where $l(a') = l - 2$, $l(b') = m - 2$. Conjugation of $ab = ba$ by x_1 yields $a'b' = b'a'$, and a', b' are powers of a common c', by induction. But then a, b are both powers of $c = x_1 c' x_1^{-1}$.

The desired conclusion holds in every case, and the proof is complete.

Proposition 6. (i) In a free group F, nth roots, when they exist, are unique, that is, if $a, b \in F$ satisfy $a^n = b^n$, $n \in \mathbb{N}$, then $a = b$.

(ii) Any element $w \in F$ has only finitely many roots, that is, the set $\{a \in F \mid a^n = w$, some $n \in \mathbb{N}\}$ is finite.

Proof. (i) Write $a = u^{-1} \check{a} u$, $b = v^{-1} \check{b} v$ as in (5), with $\check{a}$, $\check{b}$ cyclically reduced and $l(u) = r$, $l(v) = s$, say. Now apply (6) to the equations $a^n = b^n$ and $a^{2n} = b^{2n}$ to obtain

$$nl(\check{a}) + 2r = nl(\check{b}) + 2s \ ,$$

$$2nl(\check{a}) + 2r = 2nl(\check{b}) + 2s \ .$$

Thus, $l(\check{a}) = l(\check{b})$ and $r = s$, and since the equation

$$u^{-1} \check{a}^n u = v^{-1} \check{b}^n v$$

involves no cancellation, it follows that $u = v$ and $\overset{\vee}{a} = \overset{\vee}{b}$, whence $a = b$.

(ii) Let a be a root of w, say $a^n = w$, $n \in \mathbb{N}$. If $w = e$, the result follows from Proposition 5. If $w \neq e$, neither is a nor $\overset{\vee}{a}$, and it follows from (6) that $n \leq l(w)$. Then, w is an nth power for at most finitely many n and for each such n, w is the nth power of at most one element by part (i). Hence the total number of roots is finite.

Proposition 6(i) can be used to strengthen Lemma 1, as follows.

Lemma 3. If $a^h b^k = b^k a^h$ for $a,b \in F$ and $h,k \in \mathbb{Z} \setminus \{O\}$, then a and b are powers of a common element.

Proof. By simple manipulation, we may assume that $h,k \in \mathbb{N}$. Then

$$a^h = b^k a^h b^{-k} = (b^k a b^{-k})^h \ ,$$

whence $a = b^k a b^{-k}$, Proposition 6(i). Thus,

$$b^k = a b^k a^{-1} = (a b a^{-1})^k \ ,$$

and $b = a b a^{-1}$ for the same reason. Hence $ab = ba$, and the result follows from Lemma 2.

Proposition 7. Commutation is an equivalence relation on $F \setminus \{e\}$, that is, the centralizer $C(w) := \{w \in F \mid aw = wa\}$ of any $a \in F \setminus \{e\}$ is abelian.

Proof. We must show that if $u,v \in F$ both commute with some $a \in F \setminus \{e\}$, then they commute with each other. Assume that $ua = au$ and $va = av$, and that $u \neq e \neq v$ to avoid triviality. Thus, by Lemma 2, there exist $b,d \in F$ and $p,q,r,s \in \mathbb{Z} \setminus \{O\}$ such that

$$u = b^p, \ a = b^q, \ a = d^r, \ v = d^s \ .$$

But then b^q and d^r commute, as they are equal, and it follows from Lemma 3 that there is a $c \in F$ and $h,k \in \mathbb{Z}$ such that

$$b = c^h, \ d = c^k \ .$$

Thus, $u = c^{hp}$ and $v = c^{ks}$ must commute.

Finally, Proposition 7 can be strengthened as follows.

Theorem 2. For any $w \in F \setminus \{e\}$, $C(w)$ is an infinite cyclic group.

Proof. Since $e \ne w \in C(w)$, $C(w)$ cannot be finite, by Proposition 5.

Now let d be an element of minimal length in $C(w) \setminus \{e\}$, and let $v \in C(w)$ be arbitrary. Then we claim that v is a power of d. Since d and v commute (by Proposition 7), they are powers of a common a (Lemma 2):

$$d = a^h, \quad v = a^k .$$

From the second equation, a^k commutes with w, whence $a \in C(w)$ by Lemma 3. Applying (6) to the first equation,

$$l(d) = l(a^h) = |h| l(\check{a}) + 2r = (|h| - 1) l(\check{a}) + l(a) ,$$

where $\check{a}$ and r are as in (5), and $h \ne 0$ as $d \ne e$. Since $a \in C(w)$, it follows from the minimality of $l(d)$ that $|h| = 1$, and so $v = d^{\pm k}$, as required.

Exercises

1. Given a group G and a subset $X \subseteq G$, let $<X> = \cap \{H \le G \mid H \supseteq X\}$ and let W be the set of elements of G that can be written as words in $X^{\pm}$. Prove that $W = <X>$.

2. Let G and H be groups, and $\theta, \phi : G \to H$ homomorphisms, and let $X \subseteq G$ be such that $G = <X>$. Prove that if θ, ϕ agree on X (i.e. $x \theta = x \phi$, $\forall x \in X$) then $\theta = \phi$ (i.e. $g \theta = g \phi$, $\forall g \in G$).

3. Prove directly from the definition that the infinite cyclic group $Z = \{x^n \mid n \in \mathbb{Z}\}$ is free of rank 1.

4. Let F be free of rank r and G a group isomorphic to F. Show that G is free of rank r.

5. Given a group G and a normal subgroup $N \triangleleft G$ with $G/N = F$ a free group, prove that G has a subgroup C such that

$$N \cap C = \{e\}, \quad NC = G, \quad C \cong F .$$

6. How many reduced words are there of length l the free group of rank r?

7. Show that the free abelian group of rank r is isomorphic to the direct product of r copies of Z.

8. Let F be free of rank r. Show that F/F' is free abelian of rank r.

9. Prove that the free group F of finite rank r has only a finite number of normal subgroups of index $n \in \mathbb{N}$. [*Hint:* Consider homomorphisms from F onto groups G of order n.] Is the same true for arbitrary subgroups (not necessarily normal) of index n?

10. Let F be free on a subset $X \subseteq F$, and let Y be a subset of X. Prove that the subgroup $H = <Y> \leq F$ is free on the set Y.

11. Let F be free on $\{x,y\}$ and define $a_n = x^{-n}yx^n$, $n \in \mathbb{N}$. Prove that no reduced word in $a_1, ..., a_r$ ($r \in \mathbb{N}$) can be equal to e, and deduce that the group $F_r := <a_1, ..., a_r>$ is free on these generators. Show that the group $F_0 = \underset{r \in \mathbb{N}}{\bigcup} F_r = <\{x_n \,|\, n \in \mathbb{N}\}>$ is free of rank $\aleph_0$.

12. Prove that two elements a,b in a free group F are conjugate in F if and only if $\check{a}$ and $\check{b}$ (as defined in (5)) are cyclic rearrangements of one another, that is, if $\check{a} = x_1 ... x_l$, then $\check{b} = x_r ... x_l x_1 ... x_{r-1}$ for some r, $1 \leq r \leq l$.

13. Prove that, in a free group F, no non-trivial element can be conjugate to its inverse. Deduce that if $H = <a>$ is an infinite cyclic subgroup of F, then $N(H) = C(a)$, where $N(H) = \{w \in F \,|\, w^{-1}Hw = H\}$ is the normalizer of H in F.

14. Show that no word $w \neq e$ is a free group can be conjugate to a proper power of itself, namely, to w^n, where $n \in \mathbb{Z}$ and $n \neq 1$.

15. Prove that the set of reduced words of even length in $F = F(\{x,y\})$ forms a subgroup, call it H. Show that H is generated by the elements

$$a = x^2, \quad b = xy, \quad c = xy^{-1}.$$

Verify that if w is one of the 30 reduced words of length 2 in $\{a,b,c\}^{\pm}$, then the process of expressing w as a reduced word in $\{x,y\}^{\pm}$ cannot involve the cancellation of the second letter of a, b or c, nor of the first letter of a^{-1}, b^{-1}, c^{-1}. Thus show that if w is a reduced word of length l in $\{a,b,c\}^{\pm}$, then as a reduced word in $\{x,y\}^{\pm}$, its length is at least l. Deduce that H is a free group of rank 3.

16. Generalize Exercise 15 from rank 2 to rank r, that is, find a set of free generators for the group of reduced words of even length in $F(\{x_1, ..., x_r\})$.

17. For a fixed l, $r \in \mathbb{N}$, what can you say about the group generated by reduced words of length l in the free group of rank r?

18. The *centre* of any group G is defined by

$$Z(G) := \{z \in G \mid zg = gz, \; \forall g \in G\} \; .$$

Prove that the centre of every non-cyclic free group is trivial.

19. Prove that no finite group can be the union of the conjugates of a proper subgroup.

20. In the free group $F = F(\{a,b\})$ of rank 2, let

$$A = \{a_k \mid k \in \mathbb{N}\}$$

be an enumeration of those reduced words that begin and end with a power of a. Prove that if

$$B = \{b^{-k} a_k b^k \mid k \in \mathbb{N}\} \; ,$$

then the subgroup $H = \langle B \rangle$ is free on B. Deduce that $H < F$, but $\bigcup_{w \in F} H^w = F$.

CHAPTER 2

It is a classical result of Dedekind that if A is a free abelian group, then so is any subgroup of A (see Chapter 6). Our purpose in this chapter and the next is to prove the non-commutative analogue, the celebrated Nielsen-Schreier theorem. In the nonabelian case, however, the rank of the subgroup may exceed the rank of the group (see Exercises 1.11, 2.6, 2.9).

The two methods of proof, due to Nielsen (1921) and Schreier (1927), are ostensibly very different (but see the comments at the start of Chapter 3 and in the Guide to the Literature) and lay the foundations for different aspects of the subsequent development of the subject. Thus, for example, Nielsen's method leads naturally to the theory of automorphisms of free groups (where there is currently much active interest), while Schreier's method provides the key of finding presentations of subgroups (see Chapter 9).

Schreier's method will be described first for the following three reasons: it seems to exhibit a basis for the subgroup in a more explicit way, it includes the case of infinite rank, and it gives a precise formula for the rank of the subgroup in the finite case. Neilsen's method enjoys other advantages, and the restriction to the case of finite rank can be avoided.

The proof which follows is divided into six steps, and the constructions are illustrated by a concrete example at each stage. The statement of the theorem itself is postponed until step 6. Throughout, F denotes the free group $F(X)$ on the set X as constructed in Chapter 1. Recall that for $w \in F$, $l(w)$ denotes the length of w as a reduced word in $X^{\pm}$.

1.   The well-ordering of F

Definition 1. A *partial ordering* on an abstract set S is a binary relation < (less than) on the elements of S which is

(O 1) irreflexive: $\forall s \in S$, not $(s < s)$, and

(O 2) transitive: $s < t$ and $t < u => s < u$.

$<$ is called a *total ordering* if it also satisfies

(O 3) the law of trichotomy: $\forall s, t \in S$, $s < t$ or $s = t$ or $t < s$,

and a *well-ordering* if the following stronger condition holds:

(O 4) every non-empty subset of S contains a least element.

Thus, for example, the usual ordering on the positive integers is a well-ordering, while that on the real numbers is a total ordering but not a well-ordering. However, any set can be well-ordered: this is intuitively clear for finite (and even countably infinite) sets, and in general is a consequence of the Axiom of Choice.

Taking this as a starting point, we well-order the elements of F as follows: first take any well-ordering on $X^{\pm}$, and then order all elements of F first of all by length, and then lexicographically within a given length.

Definition 2. Let $<$ be a well-ordering on $X^{\pm}$, and let

$$a = x_1 \dots x_l, \quad b = y_1 \dots y_m, \quad a \neq b \ ,$$

be reduced words in F. Then write $a < b$ if either $l < m$, or $l = m$ and $x_r < y_r$ in $X^{\pm}$, where $r = \min \{i \mid x_i \neq y_i\}$.

It is an easy exercise to show that this is a well-ordering. Note that the least element of any subgroup of F is e, as this is the least element of F.

Example. Consider the free group of rank 2 with $X = \{x,y\}$, and let $X^{\pm}$ be ordered by: $x < y < x^{-1} < y^{-1}$. Then the first 24 elements of F are:

$$\left. \begin{aligned} &e < x < y < x^{-1} < y^{-1} < x^2 < xy < xy^{-1} < \\ &yx < y^2 < yx^{-1} < x^{-1}y < x^{-2} < x^{-1}y^{-1} < y^{-1}x < y^{-1}x^{-1} < \\ &y^{-2} < x^3 < x^2y < x^2y^{-1} < xyx < xy^2 < xyx^{-1} < xy^{-1}x \ . \end{aligned} \right\} \tag{1}$$

The group structure and order structure on F are related as follows.

Lemma 1. Let $w = x_1 \dots x_n$ be a reduced word in $X^{\pm 1}$ with $n > 1$ and v any element of F. Then

$$v < x_1 \dots x_{n-1} => vx_n < w \ .$$

Proof. If $l(v) < n - 1$, then $l(vx_n) < n = l(w)$ and the result is clear. We can assume tha $v = x_1 \ldots x_{r-1} \, y_r \ldots y_{n-1}$ reduced, where $1 \le r \le n - 1$ and $y_r < x_r$. If $y_{n-1} = x_n^{-1}$, the $l(vx_n) = n - 2 < l(w)$, and if not, $vx_n = x_1 \ldots x_{r-1} \, y_r \ldots y_{n-1} \, x_n$ is in reduced form, and s $vx_n < w$ in either case.

2.   The Schreier transversal

From now on, let H denote a fixed subgroup of F.

Recall that a (right) coset of H is a subset of F of the form $Hw := \{ hw \, | \, h \in H \}$, for ϵ fixed $w \in F$, and that any two cosets are equal or disjoint: given u, $v \in F$, then

$$Hu = Hv \ \text{ or } \ Hu \cap Hv = \varnothing \ .$$

The cosets of H thus partition F and, by choosing one element from each, we obtain a (right) *transversal* U for H in F. For any $w \in F$, $Hw \cap U$ thus consists of a single element, which we will denote by $\bar{w}$.

Definition 3. A subset S of F has the *Schreier property* if it contains all initial segements of all its elements, that is,

$$w = x_1 \ldots x_n \in S \Rightarrow x_1 \ldots x_{n-1} \in S \ , \tag{2}$$

where $l(w) = n \ge 1$. A *Schreier transversal* for H in F is a (right) transversal for H with the Schreier property. Note that every Schreier set, and thus every Schreier transversal, contains the empty word e.

Lemma 2. Every subgroup H of F has a Schreier transversal, for example, that obtained by choosing the least element of each right coset of H in the above ordering.

Proof. Let U consist of the least element from each right coset of H in F. Then we must show that (2) holds for U. To prove the contraposed assertion, assume that $x_1 \ldots x_{n-1} \notin U$, and let $v \in U$ be the least element of $Hx_1 \ldots x_{n-1}$. Then

$$v < x_1 \ldots x_{n-1} \Rightarrow vx_n < w, \ \text{ by Lemma 1}$$

and $Hv = Hx_1 \ldots x_{n-1} \Rightarrow Hvx_n = Hw, \ \text{ postmultiplying by } x_n \ .$

Thus, $vx_n \in Hw$ and $vx_n < w$, so that w is not least in Hw, whence $w \notin U$.

Example. Take $X = \{x,y\}$ and let F be ordered as before. For H, take the normal closure of $S = \{x^3, y^2, x^{-1}y^{-1}xy\}$. We show that $|F : H| = 6$ and simultaneously find a transversal for H in F.

First, consider the cyclic group $Z_6 = \{e,a,a^2,a^3,a^4,a^5\}$ of order 6 ($a^6 = e$), and the map $\theta : \{x,y\} \to Z_6$ with $x\,\theta = a^2$, $y\,\theta = a^3$. If θ' is the extension of θ to F (see Definition 1.1), then

$$(yx^{-1})\,\theta' = (y\,\theta)\,(x\,\theta)^{-1} = a^3 a^{-2} = a \ ,$$

$$x^3\,\theta' = y^2\,\theta' = (x^{-1}y^{-1}xy)\,\theta' = e \ .$$

The first of these equations asserts that $a \in \mathrm{Im}\,\theta'$, whence θ' is onto, and the other three that $S \subseteq \mathrm{Ker}\,\theta'$, whence $H = \bar{S} \le \mathrm{Ker}\,\theta'$, since $\mathrm{Ker}\,\theta' \lhd F$. Combining these facts and using an Isomorphism Theorem,

$$|F : H| = |F : \mathrm{Ker}\,\theta'|\ |\mathrm{Ker}\,\theta' : H| \ge |F : \mathrm{Ker}\,\theta'| = |\mathrm{Im}\,\theta'| = 6 \ .$$

On the other hand, the elements $x' = Hx$, $y' = Hy$ generate H, commute, and have orders dividing 3,2, respectively. Hence, any element of F/H is equal to one of the six words

$$e, \ x', \ x'^2, \ y', \ x'y', \ x'^2y' \ ,$$

that is, any element of F is in one of the six cosets

$$H, \ Hx, \ Hx^2, \ Hy, \ Hxy, \ Hx^2y \ .$$

But since $|F : H| \ge 6$, these six cosets must be distinct, and so

$$T = \{e, \ x, \ x^2, \ y, \ xy, \ x^2y\}$$

is a transversal for H in F.

While T clearly has the Schreier property, it is not the U found in Lemma 2. Using such rules as

$$Hx^2 = Hx^{-1}, \ Hy^{-1} = Hy, \ Hxy = Hyx \ ,$$

which follow from the definition of H, it is not hard to check that, with respect to ordering (1), the least elements of the six cosets of H are

$$U = \{e, \ x, \ x^{-1}, \ y, \ xy, \ yx^{-1}\} \tag{3}$$

3.   The Schreier generators

The next step is to find suitable generators for H. This is done in terms of the transversal U of X, the free generators X of F, and the function

$$\left.\begin{array}{c} F \to U \\ w \mapsto \overline{w} \end{array}\right\}$$

defined by $Hw \cap U = \{\overline{w}\}$. Apart from such obvious properties as

$$\overline{\overline{w}} = \overline{w},\ Hw = H\overline{w},\ \forall\, w \in F\ ,$$

$$\overline{w} = w \Leftrightarrow w \in U\ ,$$

we shall make repeated use of one further property of the bar function, derived as follows. Given $u \in U,\, x \in X^{\pm}$, it is clear that

$$Hux = H\overline{ux},\ => Hu = Hux x^{-1} = H\overline{ux}\, x^{-1}\ ,$$

from which it follows that

$$\overline{ux}\, x^{-1} = u,\ \forall\, u \in U,\ x \in X^{\pm}\ . \tag{4}$$

Note that the proof which follows invokes neither the freeness of F nor the Schreier property for U, but merely requires that $e \in U$.

Lemma 3. The elements of the set

$$A := \{ux\, \overline{ux}^{-1}\, |\, u \in U,\ x \in X^{\pm}\}$$

generate H.

Proof. Since $Hux = H\overline{ux}$, it is clear that $A \subseteq H$. Now let $h \in H$ and write

$$h = x_1 \ldots x_n,\ x_i \in X^{\pm},\ 1 \le i \le n\ ,$$

a reduced word in $X^{\pm}$, and define a sequence of elements of U inductively as follows:

$$u_1 = e,\ u_{i+1} = \overline{u_i x_i},\ 1 \le i \le n\ .$$

Now put

$$a_i = u_i x_i u_{i+1}^{-1} = u_i x_i\, \overline{u_i x_i}^{-1} \in A,\ 1 \le i \le n\ ,$$

so that

$$a_1 a_2 \ldots a_n = u_1 x_1 \ldots x_n u_{n+1}^{-1} = e h u_{n+1}^{-1} . \tag{5}$$

Since the left-hand side belongs to H, so does u_{n+1}. But u_{n+1} also belongs to U, and $H \cap U = \{e\}$. Thus, $u_{n+1} = e$ and (5) expresses h as a word in the elements of A, proving that $H = <A>$, as required.

Example. The table below has rows indexed by $u \in U$ (see (3)) and columns indexed by $x \in X^{\pm}$, and contains the generator $ux\ \overline{ux}^{-1}$ in the (u,x)–place. The entry in the

	x	y	x^{-1}	y^{-1}
e	e	e	e	y^{-2}
x	x^3	e	e	$xy^{-2}x^{-1}$
x^{-1}	e	$x^{-1}yxy^{-1}$	x^{-3}	$x^{-1}y^{-1}xy^{-1}$
y	$yxy^{-1}x^{-1}$	y^2	e	e
xy	xyx^2y^{-1}	xy^2x^{-1}	$xyx^{-1}y^{-1}$	e
yx^{-1}	e	$yx^{-1}yx$	$yx^{-2}y^{-1}x^{-1}$	$yx^{-1}y^{-1}x$

Table 1

(yx^{-1},x^{-1})-place, for example, is arrived at as follows: first form the product $ux = yx^{-2}$, then find the element of U having the same x–exponent-sum mod 3 and y–exponent-sum mod 2, namely, $\overline{ux} = xy$, and finally, enter $ux\ \overline{ux}^{-1} = yx^{-2}y^{-1}x^{-1}$. The process is in general purely mechanical, *except* for the identification of $\overline{ux}$, which is done in this case using the fact that F/H is a abelian group with generators Hx and Hy of orders 3 and 2, respectively.

4.   Decomposition of the set A

A perusal of Table 1 reveals the fact that, as a set of generators, A contains redundant elements, such as

(i)   the (xy, y^{-1})–entry, which is e, and

(ii) the (x,y^{-1})–entry, which is the inverse of the (xy,y)–entry.

It turns out, however, that these are the *only* types of redundancy, and that we can say *exactly* when each type occurs, in general. Thus, $ux\,\overline{ux}^{-1} = e$ if and only if $ux \in U$, and the non–e elements in the right half of the table are precisely the inverses of the non–e elements in the left half. This prompts the definition:

$$B = \{ux\,\overline{ux}^{-1} \mid u \in U,\ x \in X,\ ux \notin U\}\ ,$$

and the following result.

Lemma 4. The sets

$$\hat{B} = \{ux\,\overline{ux}^{-1} \mid u \in U,\ x \in X^{-1},\ ux \notin U\}$$

$$B^{-1} = \{b^{-1} \mid b \in B\}$$

coincide, and

$$A \setminus \{e\} = B \cup B^{-1}\ . \tag{6}$$

Proof. For any $u \in U,\ x \in X^{\pm 1}$,

$$(ux\,\overline{ux}^{-1})^{-1} = \overline{ux}\,x^{-1}\,u^{-1} = \overline{ux}\,x^{-1}\,\overline{\overline{ux}\,x^{-1}}^{-1}\ ,$$

by (4) above, and

$$ux \notin U \Leftrightarrow ux \neq \overline{ux} \Leftrightarrow u \neq \overline{ux}\,x^{-1} \Leftrightarrow \overline{ux}\,x^{-1} \neq \overline{ux}\,x^{-1} \Leftrightarrow \overline{ux}\,x^{-1} \notin U\ .$$

Taking $x \in X,\ X^{-1}$ in turn, it follows that

$$B^{-1} \subseteq \hat{B},\ \hat{B}^{-1} \subseteq B\ .$$

The second of these is equivalent to $\hat{B} \subseteq B^{-1}$, whence $B^{-1} = \hat{B}$. This and the definition of A immediately yields (6).

5.   Freeness of the generators B

This is the core of Schreier's proof, and is accomplished by showing that, in the process of reducing the product of two elements

$$b = ux\,\overline{ux}^{-1},\quad b' = vy\,\overline{vy}^{-1} \in B \cup B^{-1} = A \setminus \{e\} \tag{7}$$

to a reduced word in $X^{\pm 1}$, neither of the 'middle factors' (x in b and y in b') can be cancelled, except in the trivial case when $b' = b^{-1}$. This is formulated precisely as follows, and the key to the proof is the Schreier property (SP) of U.

Lemma 5. Given b and b' as in (7), the product

$$x \, \overline{ux}^{-1} \, vy \tag{8}$$

is equal to a *reduced* word xwy in $X^{\pm}$ for some $w \in F$, except in the case when

$$v = \overline{ux}, \quad y = x^{-1}, \quad u = \overline{vy} . \tag{9}$$

Proof. Let

$$\overline{ux} = x_1 \ldots x_l, \quad v = y_1 \ldots y_m$$

as reduced words in $X^{\pm}$. First observe that $y_m \neq y^{-1}$, for otherwise, $vy = y_1 \ldots y_{m-1} \in U$ by SP, whereupon $b' = e$, contrary to hypothesis. Similarly, $x_l \neq x$, for otherwise,

$$ux = \overline{ux} \, x^{-1} \, x, \quad \text{by (4),}$$

$$= \overline{x_1 \ldots x_{l-1}} \, x_l$$

$$= x_1 \ldots x_l, \quad \text{by } (SP)$$

$$= \overline{ux} \in U ,$$

which is again impossible. Thus, $x \, \overline{ux}^{-1}$ begins with x, vy ends with y, and it only remains to examine the cancellation between $\overline{ux}^{-1}$ and v.

Under this, the reduced form of the product $\overline{ux}^{-1} \, vy$ retains the final y, for otherwsie $vy = y_1 \ldots y_m y$ would be an initial segment of $\overline{ux}$, and thus in U by SP, which is not so. Similarly, the reduced form of $x \, \overline{ux}^{-1} \, v$ retains the initial x, for otherwise

$$(x \, \overline{ux}^{-1})^{-1} = \overline{ux} \, x^{-1}$$

would be an initial segment of v, and so in U by SP, whence by (4),

$$ux = \overline{ux} \, x^{-1} \, x = (\overline{ux} \, x^{-1}) x = \overline{ux}$$

is also an U, which is forbidden.

Thus, neither x nor y can cancel unless they both do (with each other), in which case

$$v = \overline{ux}, \quad y = x^{-1} \; ,$$

and (by (4)),

$$\overline{vy} = \overline{\overline{ux}\,x^{-1}} = u \; .$$

6.   Conclusion

Theorem 1 (Nielsen-Schreier). Let F be a free group and H a subgroup of F. Then H is free. Moreover, if $|F : H| = g$ and $r(F) = r$ are both finite, then

$$r(H) = (r - 1)g + 1 \; . \tag{10}$$

Proof. Let X be a set of free generators for F and U a Schreier transversal for H (Lemma 2). The resulting set A generates H (Lemma 3), and thus, so does the subset B (Lemma 4). Now let

$$w = b_1 b_2 \dots b_n, \quad b_i = u_i x_i \overline{u_i x_i}^{-1}, \quad 1 \le i \le n, \quad n \in \mathbb{N}$$

be a reduced word in $B^{\pm} = A \setminus \{e\}$, so that elements $b = ux\,\overline{ux}^{-1}$, $b^{-1} = vy\,\overline{vy}^{-1}$ satisfying (9) cannot stand in adjacent places of w in either order (note the symmetry of (9)). By Lemma 5, this means that the reduced form of w as a word in $X^{\pm}$ contains the middle parts x_i of b_i $(1 \le i \le n)$ as separate letters, and its length is thus at least n. Since $n \ge 1$, $w \ne e$. It now follows from Proposition 1.3 that H is free on B.

For the numerical part, observe that the elements of B, indexed by pairs $(u,x) \in U \times X$ with $ux \notin U$, are all distinct, by Lemma 5. It follows that

$$r(H) = gr - b \; ,$$

where

$$b = |\{(u,x,v) \in U \times X \times U \mid ux = v\}| \; . \tag{11}$$

It is therefore sufficient to prove that the right-hand side of (11) is equal to $g - 1$.

To see this, consider the graph T with g vertices labelled by elements of U, and having a directed edge from u to v, labelled x, if and only if $ux = v$. Now T is connected (by the

SP, every vertex is connected by a path to *e*) and has no circuits (since *F* is free on *X*). *T* is thus a *tree* and so has $g - 1$ edges, by Euler's formula. Since the edges of *T* are by construction in one-to-one correspondence with the members of the set on the right-hand side of (11), it follows that $b = g - 1$, as required. This completes the proof.

Example. In the case of our concrete example (see (3)), *T* looks like this:

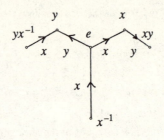

Fig. 4

and the formula gives $r(H) = 7$, as we already know from Table 1.

Now let *H* be a subgroup of infinite index in a free group *F* of rank ≥ 2. Then there is no formula for $r(H)$ in general (see Exercise 1.11). However, Schreier's method does have the following useful consequence.

Corollary 1. Let $H \leq F = F(X)$, with $|F : H| = \infty$, and suppose that *H* contains a nontrivial normal subgroup *N* of *F*. Then $r(H) = \infty$.

Proof. Let *U* be a Schreier transversal for *H* in *F*, and

$$e \neq w = x_1 \dots x_l \in N \subseteq H .$$

Then for each $u \in U$, $Huw = Huwu^{-1}u = Hu$, as $w \in N \lhd F$, so that $\overline{uw} = u \neq uw$, and $uw \notin U$. Thus, there is a least k, $1 \leq k \leq l$, such that $ux_1 \dots x_k \notin U$ and $u_k := ux_1 \dots x_{k-1} \in U$, and we have associated to each $u \in U$ a Schreier generator $u_k x_k \overline{u_k x_k}^{-1}$ of *H*. Now *k* is a function of $u \in U$ taking values in the finite set $\{1, 2, \dots, l\}$. Since $|U| = |F : H|$ is infinite, at least one of its pre-images is infinite, that is, there is an infinite subset *V* of *U* such that for all $v \in V$, the Schreier generator of *H* constructed above is $v_k x_k \overline{v_k x_k}^{-1}$, for the *same* value of *k*. If two of these are equal, say $v_k x_k \overline{v_k x_k}^{-1} = v'_k x_k \overline{v'_k x_k}^{-1}$, for $v, v' \in V$, then $v_k = v'_k$ (by Lemma 5), whence $v = v'$. This construction thus yields infinitely many distinct free generaors for *H*, which therefore must

have infinite rank.

Exercises

1. Prove that the relation $<$ of Definition 2 satisfies (01), (02), (04).

2. What is the 666th element of $F = F(\{x,y\})$ in the ordering (1)?

3. With the same F and $<$, what is the second smallest element of the commutator sub-group F'?

4. With the same F and $<$ again, let $H = \bar{S}$ with $S = \{x^4, y^2, (xy)^2\}$. Show that $|F : H| = 8$ (using D_4) and write down a Schreier transversal for H in F. Compile a table like Table 1, and thus obtain a basis B of H.

5. Prove that the permutations $x_1 = (12)$, $x_2 = (23)$, $x_3 = (34)$ generate the symmetric group S_4. Use the transversal $U = \{e, x_1\}$ of the alternating group A_4 to find Schreier generators for A_4.

6. Convince yourself that $U = \{x^k y^l \mid k, l \in \mathbb{Z}\}$ is a Schreier transversal for F' in $F = F(\{x,y\})$. Write down the corresponding Schreier generators for F' and express the commutator $x^{-2}y^{-3}x^2y^3$ in terms of them.

7. Prove that every subgroup of finite index in a finitely-generated group is finitely generated.

8. Express each of the seven free generators in the left half of Table 1 as products of conjugates of $a = x^3, b = y^2, c = x^{-1}y^{-1}xy$.

9. Prove that the free group of rank 2 contains a free subgroup of any given finite rank as a normal subgroup.

10. Prove Euler's formula for planar graphs: let Γ have v vertices, e edges (all straight line segments) and f faces (polygonal regions bounded by edges of Γ and containing no vertices of Γ in their interiors). Then $v - e + f$ is equal to the number of connected components of Γ.

11. Use the Nielsen-Schreier theorem and the result of Exercise 1.18 to give a quick proof of Theorem 1.2.

12. Let $F = F(X)$ be an arbitrary free group, and $H \leq F$ the subgroup consisting of words of even length. Use Schreier's method to obtain a basis for H (cf. Exx. 1.15, 1.16).

13. Let $F = F(X)$ be a free group of finite rank $|X| = r$, and let $n = \mathbb{N}$ be a fixed natural number. Prove that F contains only finitely many n-element Schreier sets, and that if U is one of these, the number of functions $U \times X \to U$ is finite. Deduce that F has only finitely many subgroups of index n.

14. Use the previous exercise to show that a finitely-generated group has only finitely many subgroups of a given finite index.

CHAPTER 3

The idea behind Nielsen's original proof of the subgroup theorem is as follows. Given a finite subset U of a free group F, perform certain operations on U (called Nielsen transformations) that reduce it to a set V of a special form (called Nielsen reduced.) Since it can be shown that $<U> = <V>$ and that V is a basis for $<V>$, the subgroup theorem follows in the finitely-generated case. That $<U> = <V>$ is obvious, and the freeness of $<V>$ follows from Proposition 1.3 in much the same way as the freeness of $<B>$ in Chapter 2. Another point in common with Schreier's method is the crucial role played by a well-ordering of F, which is used here to prove that any finite U can be carried by N-transformations into an N-reduced V (Theorem 1).

As in the preceding chapter, we illustrate the steps in the proof by giving a specific example, and then go on to derive a series of consequences of the method, including its extension to cover the non-finitely-generated case.

1.   The finitely-generated case

Let $F = F(X)$ be a free group and U a finite subset of F. As the words in U together involve only a finite set of generators from X, we can assume (by Exercise 1.10) that X is finite. We think of $U = (u_1, u_2, ..., u_n)$ as an ordered set and, as usual, write $l(w)$ for the length of $w \in F$ as a reduced word in $X^{\pm}$.

Definition 1. An *elementary Nielsen transformation* of U is of one of the following three types: for some i, $1 \le i \le n$,

(T0)   delete u_i, where $u_i = e$, $(\backslash i)$

(T1)   replace u_i by u_i^{-1} (i')

(T2) replace u_i by $u_i u_j$, $j \neq i$, $\hspace{4cm}$ (ij)

and leave u_k fixed for all $k \neq i$. A *Nielsen transformation* is a finite sequence of elementary Nielsen transformations, and is called *regular* if no (T0)'s are involved.

Remarks. 1. If τ is a Nielsen transformation involving k applications of (T0), then $|U\tau| = |U| - k$. For example, if $U = (a,b,e)$, and $\tau = (2')\,(12)\,(\backslash 3)$, then $U_\tau = (ab^{-1}, b^{-1})$.

2. Our definition of "elementary" is fairly minimal: many authors also include the transformations

(T2′) replace u_i by $u_i u_j^{\pm 1}$, or by $u_j^{\pm 1} u_i$, $j \neq i$, and

(T3)   interchange u_i and u_j, $i \neq j$, or

(T3′) replace u_i by $u_{i\pi}$, $1 \leq i \leq n$, for any permutation $\pi \in \mathrm{Sym}\{1,2,...,n\}$.

We leave it as an exercise to show that each of these is a regular N-transformation in the sense of Definition 1.

3. The elementary N-transformations correspond to the following elementary row operations performed as an integer matrix:

(M0) deletion of a row of zeros,

(M1) multiplying a row by -1,

(M2) adding one row to another,

in the proof of the Basis theorem (see Chapter 6). These in turn correspond to the elementary row operations of linear algebra, except that multiplication of a row by scalars other than ± 1 is not allowed, as ± 1 are the only units in $\mathbb{Z}$.

Lemma 1. (i) If U is carried into V by an N-transformation τ, then $<U> = <V>$.

(ii) The regular Nielsen transformations form a group.

Proof. (i) It is sufficient to prove that $<U_\tau> = <U>$ in the case when τ is elementary, and this is obvious.

(ii) It is sufficient to observe that (T1), (T2) have inverses, namely:

$$(i')^{-1} = (i'), \quad (ij)^{-1} = (j')\,(ij)\,(j') \ ,$$

for $1 \leq i,j \leq n$, $i \neq j$.

Our aim is now to reduce a given U, by a Nielsen transformation, to a set V such that only a limited amount of cancellation is possible between the elements $V^\pm$. This is made

precise by the following defintion.

Defintion 2. A set $V \subseteq F$ is called *Nielsen reduced* (*N*-reduced) if, for all a, b, $c \in V$, the following conditions hold:

(N1) $l(a) \neq 0$,

(N2) $a \neq b^{-1} \Rightarrow l(ab) \geq l(a), l(b)$,

(N3) $a \neq b^{-1} \neq c \Rightarrow l(abc) > l(a) - l(b) + l(c)$.

Recall the proof of associativity of $F(X)$ in Theorem 1.1. If

$$a = x_1 \ldots x_l, \quad b = y_1 \ldots y_m, \quad c = z_1 \ldots z_n ,$$

$$ab = x_1 \ldots x_{l-r} y_{r+1} \ldots y_m, \quad bc = y_1 \ldots y_{m-s} z_{s+1} \ldots z_n ,$$

all reduced, then (N1), (N2), (N3) respectively assert that $l \neq 0$, i.e. $a \neq e$; $r \leq \frac{1}{2} l, \frac{1}{2} m$, i.e. no more than half of a or b can cancel in forming ab; $r + s < m$, i.e. Case 1 of the proof that $(ab)c = a(bc)$.

The second and third of these yield parts (i) and (ii) of the following lemma, respectively, and part (iii) is a consequence of either.

Lemma 2. Let $V \subseteq F$ be *N*-reduced, and $w = v_1 \ldots v_n$ a reduced word in $V^{\pm}$. Then

(i) $l(w) \geq l(v_i)$, $1 \leq i \leq n$,

(ii) $l(w) \geq n$,

(iii) V is a basis for the subgroup $<V>$ of F.

Proof. Let $l(v_i) = l_i$, $1 \leq i \leq n$, $l(v_i v_{i+1}) = l_i + l_{i+1} - 2r_i$, $1 \leq i \leq n-1$, so that (N1), (N2), (N3) respectively assert that

$$l_i \geq 1, \quad 1 \leq i \leq n ,$$

$$r_i \leq \frac{1}{2} l_i, \frac{1}{2} l_{i+1}, \quad 1 \leq i \leq n-1 , \tag{1}$$

$$r_{i-1} + r_i < l_i, \quad 2 \leq i \leq n-1 . \tag{2}$$

Thus we can write for $2 \leq i \leq n-1$,

$$v_i = a_i \, \tilde{v}_i \, a_{i+1}^{-1} , \tag{3}$$

reduced, where $l(a_i) = r_{i-1}$, so that

$$l_i' := l(\tilde{v}_i) = l_i - r_{i-1} - r_i > 0 \ , \tag{4}$$

by (2), and

$$v_{i-1}v_i v_{i+1} = a_{i-1}\tilde{v}_{i-1}\tilde{v}_i\tilde{v}_{i+1}a_{i+1}^{-1} \tag{5}$$

is in reduced form. Not put $r_0 = r_n = 0$ and $a_1 = a_{n+1} = e$, so that (3) now defines $\tilde{v}_i$ for all i with $1 \le i \le n$. Moreover, by (1),

$$l_1' := l(\tilde{v}_1) \ge \frac{1}{2}l_1, \ \ l_n' := l(\tilde{v}_n) \ge \frac{1}{2}l_n \ ,$$

so that (4) holds for $1 \le i \le n$, and, by (1) again

$$l_1' \ge r_1, \ \ l_n' \ge r_n \ . \tag{6}$$

Now because of (5),

$$w = \tilde{v}_1 \ \tilde{v}_2 \dots \tilde{v}_n$$

is in reduced form, and part (ii) follows from (4). Furthermore,

$$v_2 \dots v_n = a_2\tilde{v}_2 \dots \tilde{v}_n, \ v_1 \dots v_{n-1} = \tilde{v}_1 \dots \tilde{v}_{n-1} a_{n-1}^{-1} \ ,$$

whence it follows from (6) that

$$l(w) \ge l(v_2 \dots v_n), \ l(v_1 \dots v_{n-1}) \ . \tag{7}$$

Repeated application of (7) now yields part (i). Either (i) or (ii) is sufficient to guarantee that

$$n \ge 1 \Rightarrow w \ne e \ ,$$

and (iii) now follows from Proposition 1.3.

Theorem 1. Let $U = (u_1, \dots, u_n)$ be an ordered n-tuple of elements of a free group $F = F(X)$, $n \in \mathbb{N}$. Then U can be carried by a Nielsen transformation into an N-reduced tuple V.

Proof. The proof proceeds by performing Nielsen transformations successively to ensure (N2), (N1), (N3), respectively. Call $\sum\limits_{i=1}^{n} l(u_i)$ the *total X-length* of U.

First suppose that (N2) fails for U, so that

$$l(u_i^\delta u_j^\varepsilon) < l(u_i^\delta) \quad \text{or} \quad l(u_j^\varepsilon), \quad u_i^\delta \neq u_j^{-\varepsilon} \, ,$$

where $1 \leq i,j \leq n$ and $\delta, \varepsilon \in \{\pm 1\}$. Since the equation $l(w^2) < l(w)$ is impossible in a free group (see formula (1.6)), $u_i^\delta \neq u_j^\varepsilon$, and it follows that $i \neq j$. Replacing U by $U(i')$, $U(j')$ or $U(i')(j')$ accordingly as $\delta = -1$, $\varepsilon = -1$ or both, we can assume that

$$l(u_i u_j) < l(u_i) \quad \text{or} \quad l(u_j) \, .$$

Replacing U by $U(ij)$ in the former case by $U(i')(j')(ji)$ in the latter, we obtain an n-tuple of shorter total X-length. Now let V be an n-tuple of minimal total X-length obtainable from U by a regular N-transformation. Then (N2) holds for V.

Now use (T0) to delete any e's occurring in V. By renaming the resulting set U, we can assume that (N2) and (N1) hold for U.

Turning to (N3), let a, b, $c \in U^\pm$ satisfying $a \neq b^{-1} \neq c$, and consider the product abc. By (N2), at most half of b cancels in forming ab, and similarly in forming bc. Thus we can write

$$a = a'p^{-1}, \, b = pb'q^{-1}, \, c = qc^{-1} \, ,$$

$$ab = a'b'q^{-1}, \quad bc = p'b'c' \, ,$$

all reduced. If $b' \neq e$, then $abc = a'b'c'$ reduced, and

$$l(abc) = l(a) - l(b) + l(c) + 2l(b') > l(a) - l(b) + l(c) \, ,$$

and (N3) holds for this triple. Now suppose that $b' = e$, so that

$$a = a'p^{-1}, \quad b = pq^{-1}, \quad c = qc' \, , \tag{8}$$

and (N3) is indeed violated. Because of (N2), however, we have

$$l(p) = l(q) \leq l(a'), \, l(c') \, . \tag{9}$$

In this case, we have the option of replacing

$$a^{-1} = pa'^{-1} \quad \text{by} \quad (ab)^{-1} = qa'^{-1}, \quad \text{or} \quad c = qc' \quad \text{by} \quad bc = pc'$$

by a transformation of type (T2) that does not alter the total X-length. To avoid the situation described by (8), (9), we have only to exercise a preference for words w such that $w^{\pm 1}$ begins with one of p or q rather than the other. This is achieved by ordering the doubletons

$\{w^{\pm}\} \in F \setminus E$ using a symmetrised version of the ordering of F defined in the previous chapter.

First, let $<$ be a well-ordering of F that respects length and is lexicographical within a given length (as in Section 2.1). Now define the left half of $w \in F \setminus E$ to be its initial sub-word of $L(w)$ of length $\left[\dfrac{l(w) + 1}{2} \right]$, where [ ] denotes the integer part function. Finally, order the pairs $w^{\pm 1}$ in $F \setminus E$ by declaring that $w_1^{\pm 1} \ll w_2^{\pm 1}$ if and only if

either a) $l(w_1) < l(w_2)$, or these are equal and either

b) $\min \{L(w_1), L(w_1^{-1})\} < \min \{L(w_2), L(w_2^{-1})\}$,

or c) these mins are equal, and $\max \{L(w_1), L(w_1^{-1})\} < \max \{L(w_2), L(w_2^{-1})\}$.

Now assume that a, b, $c \in U^{\pm}$ violate (N3), so that (8) and (9) hold. Then

$$p < q \Rightarrow bc = pc' \ll qc' = c, \text{ and} \left. \atop q < p \Rightarrow ab = aq^{-1} \ll a'p^{-1} = a , \right\} \tag{10}$$

by (9) and the definition of $\ll$. (For the sake of clarity, we drop the superscript $\pm$ from U and its elements). In either case, there is an (T2$'$)-transformation of U which

a)   fixes the total X-length,

b)   replaces one member of U by a smaller element of F

c)   fixes the remaining members of U,

so that $U \tau \ll U$ in the obvious lexicographical sense.

Finally, let $\mathcal{U}$ denote the family of all sets obtainable from U by finite sequences of (T2$'$) transformations that fix the total X-length. Since every member of $\mathcal{U}$ is a $|U|$-tuple of words of bounded length in a finite alphabet, $\mathcal{U}$ is a finite set. But if $\mathcal{U}$ contained no set satisfying (N3), the process of the previous paragraph would yield an infinite descending chain in $\mathcal{U}$. Thus $\mathcal{U}$ contains a set V for which (N3) holds. Since V also satisfies (N1) and (N2), the proof is complete.

Corollary 1. Every finitely-generated subgroup H of a free group F is free.

Proof. Let U be a finite set generating H and τ a Nielsen transformation such that $V = U\tau$ is N-reduced, which exists by Theorem 1. Then $H = \langle V \rangle$ by Lemma 1, and V is a basis for

H by Lemma 2. Hence, H is free of rank $|V|$.

2.   Example 1

Let H be the subgroup of $F = F(\{x,y\})$ consisting of all words of even length (cf. Exercise 1.15). Clearly, H is generated by the 12 words of length 2 in F, and thus by the 6 pairwise non-inverse components of

$$U = (u_1, ..., u_6) = (x^2, xy, xy^{-1}, yx, y^2, y^{-1}x) \ .$$

Then with the above notation,

$$U(35)(6')(46)(6')(1')(61)(1') = (x^2, xy, xy, y^2, y^2, y^{-1}, x^{-1}) = U_1, \ \text{say}$$

$$U_1(2')(32)(2')(4')(54)(4')(62) = (x^2, xy, e, y^2, e, e) = U_2, \ \text{say}$$

$$\text{and } U_2(\backslash 3)(\backslash 5)(\backslash 6) = (x^2, xy, y^2) = U_3 = (v_1, v_2, v_3), \ \text{say} \ .$$

Of the 36 words ab of length 2 in $U_3^{\pm}$, 6 are not reduced and 6 are squares. Of the remaining 24, only 4 involve any cancellation, namely $(v_1^{-1}v_2)^{\pm}$ and $(v_2v_3^{-1})^{\pm 1}$, and these all have length 2. Thus, (N2) holds for U_3 and so, clearly, does (N1). (Since $l(h) \geq 2$ for $h \in H \setminus E$ and $r(H) = 3$ by Theorem 2.1, U_3 also achieves the minimal total X-length of 6).

(N3), however, is violated, since

$$l(v_3 v_2^{-1} v_1) = 2 = l(v_3) - l(v_2^{-1}) + l(v_1) \ .$$

Rather than apply (T2')-transformations at random to U_3, let us determine the lexicographically smallest V in the ordering $\ll$ induced by $x < y < x^{-1} < y^{-1}$ (see (2.1)). The first 6 words of length 2 under $<$ contain an inverse pair. Omitting the second of these (yx^{-1}) and including the next $(x^{-1}y)$, we obtain column 2 in the following table.

<	a	a^{-1}	$L(a)$	$L(a^{-1})$	≪
1	x^2	x^{-2}	x	x^{-1}	2
2	xy	$y^{-1}x^{-1}$	x	y^{-1}	3
3	xy^{-1}	yx^{-1}	x	y	1
4	yx	$x^{-1}y^{-1}$	y	x^{-1}	4
5	y^2	y^{-2}	y	y^{-1}	5
6	$x^{-1}y$	$y^{-1}x$	x^{-1}	y^{-1}	6

Table 2

The first column gives the position in the ordering < . Noting that min $\{L(a), L(a^{-1})\}$ turns out to be just $L(a)$ in every case (column 4), the positions in the ordering ≪ are as given in the last column. Thus, the smallest triple of all under ≪ is $V = (xy^{-1}, x^2, xy)$, and since V is just a permutation of $U_3(2')(32)(2')(3')$, it is the desired basis of H.

3. The general case

Let F be any free group and H any subgroup of F. We use the ordering ≪ of §1 to obtain a basis for H in the following way. Given $w \in H$, define

$$H(w) := <\{h \in H \mid h^{\pm 1} \ll w^{\pm 1}\}> ,$$

and then put

$$A := \{a \in H \mid a \notin H(a)\} ,$$

so that A consists of just those $a \in H$ that cannot be written as words in smaller elements of H. Now let B be a subset of A obtained by throwing away exactly one of each inverse pair of elements of A, so that $A = B \cup B^{-1}$.

Corollary 2. The set B just defined is a basis for H.

Proof. By Lemma 2, it is sufficient to prove that B generates H and satisfies (N1), (N2) and (N3).

$< B > = H$: if not, let $w^{\pm}$ be the least among the inverse pairs in $H \setminus < B >$ (see Exercise 5). Then $w \notin A = B^{\pm}$, and so $w \in H(w)$. But by minimality of $w^{\pm 1}$, every h with $h^{\pm 1} \ll w^{\pm 1}$ lies in $< B >$, whence $H(w) \subseteq < B >$. It follows that $w \in < B >$ and this is a contradiction.

(N1): since $e \in H(e) = < \varnothing > = E$, $e \notin B$.

(N2): Let $x, y \in B$, $x \neq y$, and consider the word

$$z = (x^{\gamma} y^{\delta})^{\varepsilon}, \quad \gamma, \delta, \varepsilon \in \{\pm 1\} .$$

We claim that $z^{\pm 1} \gg x^{\pm 1}, y^{\pm 1}$. To prove this, assume that $x^{\pm 1} \ll y^{\pm 1}$ (wihtout loss of generality) and that $z^{\pm 1} \ll y^{\pm 1}$. Whatever the values of $\gamma, \delta, \varepsilon$, it is clear that $y \in < x, z > \leq H(y)$, contradicting the fact that $y \in B$. Since $\ll$ respects length, $l(z) \geq l(x), l(y)$ and (N2) follows.

(N3): if (N3) were violated in B, then we would have the situation (10) described in the proof of Theorem 1, and there would be an (N2')-transformation of B fixing all but one of its elements and replacing that element, y say, by a smaller one, z say; $z \ll y$. But it is shown in the previous paragraph that any such $z \gg y$. So this situation cannot arise, and (N3) holds for B.

4. Further applications

We now derive an interrelated sequence of consequences of Nielsen's method of which the main results appear in Corollaries 5, 7 and 10.

Consider the following two questions: which groups can be isomorphic to a proper subgroup (a proper factor group) of themselves? Certainly not any finite groups. On the other hand, the infinite cyclic group is isomorphic to every non-trivial subgroup of itself, although not to any proper factor group, since these are all finite. Again, the free group $F(\{x, y\})$ is isomorphic to $< y, x^{-1} y x >$ (see Exercise 1.11), which is clearly a proper subgroup, and there is a similar construction for the free group of any given finite rank. Finally, consider the free group $F = F(X)$, where $X = \{x_n \mid n \in \mathbb{N}\}$ is countably infinite. The subgroup $H = < \{x_n \mid n \geq 2\} >$ is proper, as $x_1 \notin H$, and isomorphic to F (Exercise 1.10). But also, the homorphism $\phi : F \to H$ sending x_i to $x_i (i \geq 2)$ and x_1 to e is clearly onto though not one-to-one, as $x_1 \in \mathrm{Ker} \, \phi$. Hence

$$F \cong H = \mathrm{Im} \, \phi \cong F / \mathrm{Ker} \, \phi, \quad \mathrm{Ker} \, \phi \neq E , \tag{11}$$

and F is isomorphic to a proper factor group of itself. Our first aim is to show that finitely-generated free groups do not exhibit such pathological behaviour.

Definition 3. A group is called *Hopfian* if it is not isomorphic to any proper factor group of itself.

Now let $F = F(X)$ denote a free group of rank $r = r(F)$, and consider a finite set U of generators of F. By Theorem 1, there is a Nielsen transformation τ such that $V := U\tau$ is N-reduced. By Lemmas 1 and 2, V is a basis for F, and so $|V| = r$, by Proposition 1.1. Now the cardinality of a set is fixed under transformations of types (N1) and (N2), and diminished by one for each of type (N0). It follows that

$$r = |V| = |U\tau| \le |U| \, , \tag{12}$$

and we have the following result.

Corollary 3. A free group of finite rank r cannot be generated by fewer than r elements.

It follows from the foregoing that equality can occur in (12) only when τ involves no (N0), that is, τ is regular and so has an inverse, by Lemma 1(ii). It is not hard to show that (N1) and (N2) carry bases into bases, and so equality in (12) ensures that $U = V\tau^{-1}$ is a basis for F.

Corollary 4. If a free group F of finite rank r is generated by a set U of r elements, then U is a basis for F.

Now let $F = F(X)$, with $|X| = r < \infty$, and $N \lhd F$ such that $F/N \cong F$, so that the composite

$$\phi : F \xrightarrow{\text{nat}} F/N \cong F$$

is an epimorphism. Now let $w = w(X)$ be a reduced word in Ker ϕ. Then

$$e = w\phi = w(X)\phi = w(X\phi) \tag{13}$$

is a reduced word in $X\phi = \{x\phi \,|\, x \in X\}$. But $X\phi$ generates F, and so has r elements (Corollary 3) and is thus a basis (Corollary 4). It follows from (13) that w is the empty word, whence Ker $\phi = N$ is trivial.

Corollary 5. Finitely-generated free groups are Hopfian.

Definition 4. Let X be a class of groups. Then a group G is said to be *locally X* if every finitely generated subgroup of G belongs to X.

We shall now use the previous result to obtain an example of a group that is locally free but not free.

Example 2. Let $U = \bigcup_{n \in \mathbb{N}} F_n$, where each F_n is a free group of rank $\leq r$, for some $r \geq 0$, and $F_n \subseteq F_{n+1}$ for all $n \in \mathbb{N}$. Then we know that

$$\text{each } F_n \text{ is Hopfian (Corollary 5), and} \tag{14}$$

$$\text{every finitely-generated subgroup of } U \text{ is free (Corollary 1) .} \tag{15}$$

We claim that U is Hopfian. For if not, let r be least for which this is false. Then, for this r, U has a normal subgroup $N \neq E$ such that

$$U \cong U/N, \text{ and} \tag{16}$$

$$N \cap F_n \neq E \text{ for large enough } n . \tag{17}$$

Then

$$U/N = \bigcup_{n \in \mathbb{N}} F_n N/N, \text{ and } F_n N/N \text{ is free, by (15) and (16) ,}$$

but, for large enough n,

$$F_n N/N \cong F_n/F_n \cap N \text{ has rank } \leq r, \text{ by (14) and (17) .}$$

It follows that U is a union of free groups of rank $\leq r - 1$, and thus is Hopfian, by minimality of r. This contradiction proves the claim.

Thus, U is a Hopfian group which is at most countably generated, and so cannot be free of rank $\aleph_0$. If, on the other hand, $U = \bigcup_{n \in \mathbb{N}} F_n$ is a *proper* union (as in the first example in §9.7, U is not finitely generated. It follows that such a U cannot be free.

For the next type of application, consider the following question: given a Nielsen-reduced set V in $F(X)$, when can a member of $X^{\pm}$ belong to $<V>$? Suppose we have an equation of the form

$$x = v_1 \dots v_k ,$$

where $x \in X^{\pm}$ and the right-hand side is a reduced word in $V^{\pm}$. Then it follows from Lemma 2(ii) that

$$1 = l(x) = l(v_1 \dots v_k) \geq k .$$

Since $x \neq e$, $k \neq 0$, and so $k = 1$, that is, $x = v_1 \in V^{\pm}$. Thus, $X^{\pm} \cap <V> \subseteq V^{\pm}$, which is the only non-trivial part of the next result.

Corollary 6. If $V \subseteq F(X)$ is N-reduced, then $X^{\pm} \cap <V> = X^{\pm} \cap V^{\pm}$. If in addition $<V> = F$, then $X^{\pm} = V^{\pm}$.

Now let $\phi : F \to F$ be an automorphism of the free group $F = F(X)$, where $|X| = r < \infty$. Now let τ be a Nielsen transformation such that $(X \phi) \tau =: V$ is N-reduced. Since V generates F, it follows from Corollary 6 that $V^{\pm} = X^{\pm}$, so that $V \sigma = X$, where σ is a finite compositie of (N1)'s. Since $(X \phi) \tau \sigma = X$ and $|X \phi| = |X| = r$, $\tau \sigma$ is regular, and so has an inverse. Now ϕ is determined by $X \phi = X \sigma^{-1} \tau^{-1}$, and $\sigma^{-1} \tau^{-1}$ is made up of (N1)'s and (N2)'s, which are respectively r and $r(r-1)$ in number. These r^2 transformations thus generate Aut F.

Corollary 7. If F is a finitely-generated free group, then Aut F is finitely generated.

For our third and final application, we consider residual properties of groups. Let X be a class of groups, such finite groups ($\mathcal{F}$), free groups, abelian groups ($\mathcal{A}$), or nilpotent groups ($\mathcal{N}$), for example. It is traditional to refer to X as a property of groups:

$$G \text{ has property } X \Leftrightarrow G \text{ belongs to class } X .$$

Given one property, we can pass to a "derived" property in the following way.

Definition 5. Let X be a class of groups. Then a group G is said to be *residually-X* if, for any $g \in G \setminus E$, there is a group $H \in X$ and an epimorphism $\phi : G \twoheadrightarrow H$ such that $g \phi \neq e$. Aliter:

$$\forall g \in G \setminus E, \ \exists N \triangleleft G \text{ such that } g \notin N \text{ and } G/N \in X .$$

The class of residually-X groups is often denoted by $_R X$.

We begin to work towards residual properties of finitely-generated free groups, although the next two results hold for $F = F(X)$ with $|X|$ arbitrary.

Definition 6. An element of a free group is called *primitive* if it is a member of some basis.

Corollary 8. Let $H \leq F(X)$ and $w \in H$ a word of a minimal positive X-length in H. Then w is a primitive element of H.

Proof. Let B be a basis of H and write $w = b_1 \ldots b_k$ as a reduced word in $B^{\pm}$. Let $U = \{b_1^{\pm 1}, \ldots, b_k^{\pm 1}\} \cap B$ and $V = U \tau$ be N-reduced, where τ is a regular Nielsen transformation of U. Then $B' = V \cup (B \setminus U)$ is again a basis for H, and we can write $w = v_1 \ldots v_m$ as a reduced word in $V^{\pm}$.

Now if $m \geq 3$, it follows from (7) and the minimality of $l(w)$ that

$$l(w) \geq l(v_1 v_2 v_3), \; l(v_1), \; l(v_2), l(v_3) \geq l(w) \; .$$

But then these are all equal, and so

$$l(v_1 v_2 v_3) = l(v_1) - l(v_2) + l(v_3) \; .$$

contradicting (N3).

Thus, $m = 1$ or 2, and either

$$w = v_1 \in B'^{\pm} \text{ or } w = v_1 v_2 \text{ and } l(w) = l(v_1) = l(v_2) \; .$$

In the former case, $w \in B'$ or B'^{-1} (both of which are bases), and in the latter that $w \in B' \tau$ where τ is of type (T2$'$) (since we must have $v_1 \neq v_2$). In either case, w is primitive and the proof is complete.

A subgroup H of a group G is normal if $x^{-1} H x = H$ for all $x \in G$, that is, H is preserved by all inner automorphisms of G. We now define a stronger property.

Definition 7. A subgroup H of a group G is called *characteristic* if it is preserved by all automorphisms of G, that is, $H\phi = H$ for all $\phi \in \text{Aut } G$.

Now let $K < H \leq F(X)$, where K is a proper characteristic subgroup of H, and let w be of minimal positive X-length in H, so that $w \in$ some basis B of H, by Corollary 8. If $w \in K$, then so is every $b \in B$, since the transposition $w \longleftrightarrow b$ is an automorphism of H and K is characteristic. But then $H \leq K$, as B generates H, and this contradicts the assumption that $K < H$. Hence $w \notin K$. It follows that if

$$F = F(X) = H_1 > H_2 > \ldots > H_n > \ldots \tag{18}$$

is an infinite strictly descending chain of subgroups of F, each characteristic in its predecessor, and $l_i = \min \{l(w) | e \neq w \in H_i\}$, then $1 = l_1 < l_2 < \ldots < l_n < \ldots$. Hence, if $v \in F$ has length $k \geq 1$, then $V \notin H_{k+1}$, and we have proved the following result.

Corollary 9. Given an infinite strictly descending chain (18) of subgroups of F, each characteristic in its predecessor, then $\bigcap_{n \in \mathbb{N}} H_n = E$.

Our goal is now in sight. For any group G, write

$$G^2 = <\{g^2 \mid g \in G\}> \ ,$$

so that G^2 is a characteristic subgroup of G (as the set $\{g^2 \mid g \in G\}$ is clearly preserved by any $\phi \in \text{Aut } G$). In particular, $G^2 \lhd G$, and the square of every element of G/G^2 is the identity. G/G^2 is thus abelian, and is finite if G is finitely-generated.

Now let $F = F(X)$ with $|X| = r < \infty$, and define a chain of subgroups of F as follows:

$$H_1 = F, \ \ H_{k+1} = H_k^2, \ \ k \in \mathbb{N} \ .$$

Putting $r_k = r(H_k)$, $k \geq 1$, it is not hard to show that

$$|H_k : H_{k+1}| = 2^{r_k} \ ,$$

whence it follows from the numerical part of the Nielsen-Schreier theorem that

$$r_{k+1} = (r_k - 1)2^{r_k} + 1 \ .$$

Thus the r_k are all finite, and the F/H_k are all finite groups of 2-power order. Assuming $r \geq 1$ to avoid triviality, the chain $\{H_k > H_{k+1} \mid k \in \mathbb{N}\}$ thus satisfies all the conditions of Corollary 9. It follows that, for any $w \in F \setminus E$, $w \notin H_n$ for some $n \in \mathbb{N}$, and we have the following result.

Corollary 10. Every finitely-generated free group F is residually a finite 2-group. In particular, $F \in_R \mathcal{F}$ and $F \in_R \mathcal{N}$.

Exercises

1. By expressing the transposition: $(u_1, ..., u_i, ..., u_j, ..., u_n) \mapsto (u_1, ..., u_j, ..., u_i, ..., u_n)$ as a product of six elementary Nielsen transformations, prove that any permutation of U can be effected by a regular Nielsen transformation.

2. Let τ be the transformation of U sending u_i to $(u_i^\gamma u_j^\delta)^\varepsilon$, where $i \neq j$ and $\gamma, \delta, \varepsilon \in \{\pm 1\}$, and fixing all other u_k. Show that τ is a regular Nielsen transformation.

3. Prove that, except in the trivial case $U = \{e\}$, (N3) => (N1).

4. Let $U = (u_1, ..., u_n)$ be a basis for a subgroup H of F, and τ a regular Nielsen transformation of U. Prove that $U\tau$ is a basis for H. (Hint: for $w \in H$, let $l(w), l_\tau(w)$ denote the length of w as a reduced word in U, $U\tau$, respectively. Then prove that $l_\tau(w) \geq l(w)/k$, when τ is an elementary Nielsen transformation of type (Tk), $k = 1,2$).

5. Prove that the ordering $\ll$ in the proof of Theorem 1 is a well-ordering on the pairs $w^{\pm 1}$ in $F \setminus E$.

6. Let H be the subgroup $\overline{\{x^3, y^2, x^{-1}y^{-1}xy\}}$ of $F\{(x,y)\}$ studied in Chapter 2. Find Schreier generators B of H using the Schreier transversal $\{e, x, x^2, y, xy, x^2y\}$. Describe a Nielsen transformation τ such that $B\tau$ is N-reduced.

7. Let U be a Schreier transversal for a subgroup H of $F(X)$ such that u is of minimal length in Hu, for all $u \in U$. Prove that the Schreier basis B is N-reduced.

8. Use Exercise 1.18, and Corollaries 3.1, 3.3 to give a quick proof of Theorem 1.2.

9. Let $\phi : F(X) \twoheadrightarrow F(Y)$ be an epimorphism of free groups with $|X|$ finite. Prove that $|Y|$ is finite. Prove that $F(X)$ has a basis $V = V_1 \cup V_2$ such that ϕ maps V_1 one-to-one onto a basis for $F(Y)$ and V_2 into E. (*Hint:* Let τ be a Nielsen transformation carrying $X\phi$ into Y, and examine the effect of τ on X).

10. Prove that every finitely-generated free group F satisfies the ascending chain condition for normal subgroups, that is, if

$$H_1 \subseteq H_2 \subseteq ... \subseteq H_n \subseteq H_{n+1} \subseteq ...$$

is a chain of normal subgroups of F, then $H_i = H_j$ for all $i, j \geq$ some $N \in \mathbb{N}$.

11. Let ϕ be an automorphism of $F(\{x,y\})$ and let $c = x^{-1}y^{-1}xy$. Prove that $c\phi = w^{-1}c^{\pm 1}w$ for some $w \in F$.

12. Let F be free of finite rank r. Then it was shown in deriving Corollary 7 that Aut F is generated by r^2 elements. Can you improve on this?

13. Prove that

    (i) $x^5y^{-1}x^8$, (ii) $xyxyx$

    are primitive elements of $F(\{x,y\})$, but that

    (iii) x^2,   (iv) $x^{-1}y^{-1}xy$

    are not.

14. Use Corollary 10 to prove that, for any set X, $F(X)$ is residually finite.

15. Let H be a subgroup of finite index in a group G. Prove that H is residually finite if and only if G is.

CHAPTER 4

FREE PRESENTATIONS OF GROUPS

1. Basic concepts

Suppose that

X is a set,

$F = F(X)$ is the free group on X,

R is a subset of F,

$N = \bar{R}$ is the normal closure of R in F, and

G is the factor group F/N.

Definition 1. With this notation, we write $G = <X \mid R>$ and call this a *free presentation*, or simply a *presentation* of G. The elements of X are called *generators* and those of R *defining relators*. A group G is called *finitely presented* if it has a presentation with both X and R finite sets.

Remarks. 1. This makes precise the notion that the $x \in X$ generate G, that the $r \in R$ are equal to e in G, and that G is the "largest" group with these properties. Note the abuse of notation in referring to x and r as elements of G. This is done for convenience. It is always clear from the context to which group (F or G) a given word in $X^{\pm}$ belongs.

2. It is sometimes convenient to replace R in $<X \mid R>$ by the set of equations $R = e$, that is $\{r = e \mid r \in R\}$, called *defining relations* for G. A defining relation may even take the form "$u = v$", where $u, v \in F(X)$, corresponding to the defining relator uv^{-1}.

3. On the one hand, presentations may be looked upon as a convenient shorthand for specifying a particular group, and in the next chapter, we shall be concerned with the problem of constructing presentations of groups described in some other way. On the other hand, groups frequently "occur in nature" in the form of presentations, and this gives rise to the

harder problem of deducing properties of a group from a presentation of it. There are a number of techniques for doing this (of which none is definitive), and some of these will be described later.

Examples. 1. $<X \mid >$ is a presentation of the free group of rank $\mid X \mid$.

2. The procedure in the example in §2.2 shows that

$$<x,y \mid x^3, y^2, x^{-1}y^{-1}xy> \tag{1}$$

is a presentation for the cyclic group of order 6.

3. Similarly, Exercise 2.4 asks for a proof that

$$<x,y \mid x^4, y^2, (xy)^2>$$

is a presentation for the dihedral group of order 8.

4. By definition, every cyclic group is a homomorphic image of $Z = <x \mid >$. The corresponding kernel N is again cyclic (from elementary group theory, or by putting $r = 1$ in Theorem 2.1), and is thus either trivial or the normal closure of x^n, $n \in \mathbb{N}$. A complete list of cyclic groups is thus

$$Z = <x \mid >, \; Z_n = <x \mid x^n>, \; n \in \mathbb{N} \; . \tag{2}$$

Proposition 1. Every group has a presentation, and every finite group is finitely presented.

Proof. Let G be any group, $x \subseteq G$ a set of generators for G (take $X = G$ itself, for example), and $\theta' : F(X) \rightarrow G$ the homomorphism of Proposition 1.4. Then $G = <X \mid \text{Ker } \theta'>$. If G is finite (of order g, say), then so is $X(\mid X \mid = r$ say). Then Ker θ' is generated by a set B of cardinality $(r - 1) g + 1$, by the Nielsen-Schreier theorem. Since $<B> = \text{Ker } \theta' \lhd F(X)$, $<B> = \bar{B}$, and so $G = <X \mid B>$, which is a finite presentation.

2. Induced homomorphisms

Lemma 1. Let F, G, H be groups and $v : F \rightarrow G$, $\alpha : F \rightarrow H$ homomorphisms such that (i) Im $v = G$, and (ii) Ker $v \subseteq$ Ker α. Then there is a homomorphism $\alpha' : G \rightarrow H$ such that $v \alpha' = \alpha$.

Proof. We want to prove the existence of $\alpha' \in \text{Hom}(G,H)$ such that the diagram (Fig. 5)

Fig. 5

commutes. Given $g \in G$, pick $f \in F$ such that $fv = g$ and define $g\alpha' = f\alpha$. Such an f exists because of (i) and α' is well-defined because of (ii):

$$f'v = fv \Leftrightarrow f'f^{-1} \in \text{Ker } v$$

$$\Rightarrow f'f^{-1} \in \text{Ker } \alpha, \text{ by (ii) },$$

$$\Rightarrow f'\alpha = f\alpha .$$

The value of $f\alpha$ is thus independent of the choice of $f \in v^{-1}(g)$.

Now let $g, g' \in G$ and $f, f' \in F$ such that $fv = g, f'v = g'$. Since v is a homomorphism, $(ff')v = gg'$, and

$$(gg')\alpha' = (ff')\,\alpha, \text{ by definition of } \alpha' ,$$

$$= (f\,\alpha)\,(f'\,\alpha), \text{ as } \alpha \text{ is a homomorphism },$$

$$= (g\,\alpha')\,(g'\,\alpha') ,$$

proving that α' is a homomorphism.

Finally, $f \in v^{-1}(fv)$ for each $f \in F$, so that the value of α' on fv is just $f\alpha$. This shows that $v\,\alpha' = \alpha$ and the proof is complete.

As immediate consequences of this, we obtain the results which respectively describe factor groups and homomorphisms in the context of presentations.

Proposition 2 (von Dyck). If $G = <X\,|\,R>$ and $H = <X\,|\,S>$, where $R \subseteq S \subseteq F(X)$, then there is an epimorphism $\phi : G \to H$ fixing every $x \in X$ and such that $\text{Ker } \phi = \overline{S \setminus R}$. Conversely, every factor group of $G = <X\,|\,R>$ has a presentation $<X\,|\,S>$ with $S \supseteq R$.

Proof. Apply the lemma with v and α the natural maps. Since v is onto and $\text{Ker } v = \overline{R} \subseteq \overline{S} = \text{Ker } \alpha$, the α' exists. Then $\phi = \alpha'$ fixes every $x \in X$ (since v and α do), ϕ is

onto as $\alpha = v\phi$ is, and

$$\text{Ker } \phi = (\text{Ker } \alpha) v = \bar{S}v = \overline{S \setminus R} ,$$

as $Rv = e$ in G. For the converse, if H is a factor group of G, let ψ be the composite

$$F(X) \to G \to H$$

of natural maps, so that $R \subseteq \text{Ker } \psi$ and $H = <X \mid \text{Ker } \psi>$.

Proposition 3 (Substitution Test). Suppose we are given a presentation $G = <X \mid R>$, a group H, and a mapping $\theta : X \to H$. Then θ extends to a homomorphism $\theta'' : G \to H$ if and only if, for all $x \in X$ and all $r \in R$, the result of substituting $x \theta$ for x in r yields the identity of H.

Proof. Let $G = <X \mid >$ and consider the commutative diagram in Fig. 6:

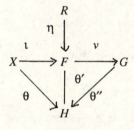

Fig. 6

where η and ι are inclusions and v is the natural map. Since F is free on X, θ extends to a unique $\theta' : F \to H$, and the substitution condition can be rephrased simply as: $R \subseteq \text{Ker } \theta'$. Since $\text{Ker } \theta' \triangleleft F$ and $\bar{R} = \text{Ker } v$, this condition is equivalent to: $\text{Ker } v \leq \text{Ker } \theta'$. The existence of a θ'' induced by θ' is now a consequence of Lemma 1. For the converse, the existence of such a θ'' entails that: $R \subseteq \bar{R} = \text{Ker } v \subseteq \text{Ker } v\theta'' = \text{Ker } \theta'$.

Remark 4. Note that when such a θ'' exists, it must be unique, since X generates G. Moreover, θ'' is an epimorphism if and only if $<X \theta> = H$.

3. Direct products

Note that Z_6 appears twice in the examples in §1: as (1) and as (2) with $n = 6$. The latter is the standard presentation for a cyclic group, and the former is the standard presentation for $Z_3 \times Z_2$ (which is isomorphic to Z_6 since 3 and 2 are coprime: see Example 7). This last observation is a consequence of the following general result.

Proposition 4. If G,H are groups presented by $<X \mid R>$, $<X \mid S>$ respectively, then their direct product $G \times H$ has the presentation

$$<X, Y \mid R, S < [X, Y]> , \tag{3}$$

where $[X,Y]$ denotes the set of commutators $\{x^{-1}y^{-1}xy \mid x \in X,\ y \in Y\}$.

Proof. Let D be the group presented by (3). By the substitution test, the "inclusions"

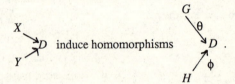

Fig. 7

(The inverted commas are used in deference to the tradition that inclusions be one-to-one). The relators $[X,Y]$ guarantee that the images of θ and ϕ centralise one another in D, and we obtain a homomorphism

$$\left. \begin{array}{l} \alpha: \ G \times H \to D \\ \quad (g,h) \mapsto g\theta h\phi \end{array} \right\} ,$$

sending (x,e) to x and (e,y) to y, for all $x \in X, y \in Y$.

On the other hand, the mapping of $X \cup Y$ into $G \times H$ sending x to (x,e) and y to (e,y) extends (by the substitution test again) to a homomorphism $\beta: D \to G \times H$. Since $\alpha\beta$ and $\beta\alpha$ both fix generating sets (in $G \times H$ and D, respectively), it follows that α and β are mutually inverse, whence α is an isomorphism.

4.   Tietze transformations

Another simple application of the substitution test provides the theoretical basis for a very important practical tool: a formal method for passing from a given presentation of a group to another presentation of the same group.

Proposition 5. Let $F = <X \mid >$, $G = <X \mid R>$ and suppose that $w, r \in F$ with w arbitrary and $r \in \overline{R} \setminus R$. If y is a symbol not in X, then both the "inclusions"

$$X \to <X \mid R, r>$$
$$X \to <X, y \mid R, y^{-1}w>$$

extend to isomorphisms with domain G.

Proof. That these mappings extend to homomorphisms of G follows at once from Proposition 3, since in each case the set R appears in the defining relators of the codomain. To obtain their inverses, observe that the maps

$$X \to <X \mid R>, \ X \cup \{y\} \to <X \mid R> \ ,$$

which fix X elementarise and send y to $w \in G$, likewise extend to homomorphisms, whose composites with the original in either direction yield identity maps, since they fix a set of generators in each case.

The four isomorphisms of Proposition 5 yield four ways of adjusting a given presentation $<X \mid R>$ to obtain another, $<X' \mid R'>$ say, of the same group. These are called *Tietze transformations* and are defined as follows, where $F = <X \mid >$ throughout.

$R +$, adjoining a relator:

$$X' = X, \ R' = R \cup \{r\} \ ,$$

where $r \in \overline{R} \setminus R$ (normal closure in F).

$R -$, removing a relator:

$$X' = X, \ R' = R \setminus \{r\} \ ,$$

where $r \in R \cap \overline{R \setminus \{r\}}$.

$X +$, adjoining a generator:

$$X' = X \cup \{y\}, \quad R' = R \cup \{y^{-1}w\},$$

where $y \notin X$ and $w \in F$.

$X-$, removing a generator:

$$X' = X \setminus \{y\}, \quad R' = R \setminus \{y^{-1}w\},$$

where $y \in X$, $w \in \langle X \setminus \{y\} \mid \rangle$ and $y^{-1}w$ is the only member of R involving y.

Example 5. Consider the von Dyck group

$$D(l,m,n) = \langle x,y \mid x^l, y^m, (xy)^n \rangle,$$

where l,m,n are integers (usually positive). We apply a succession of Tietze transformations to this presentation in accordance with the scheme set out in Fig. 8. The generators are omitted since at each stage they are just the letters involved in the relators.

	x^l	y^n	$(xy)^n$							
$X+$	x^l	y^m	$(xy)^n$	$a^{-1}xy$						
$R+$	x^l	y^m	$(xy)^n$	$a^{-1}xy$	a^n					
$R-$	x^l	y^m		$a^{-1}xy$	a^n					
$R+$	x^l	y^m		$a^{-1}xy$	a^n	$(ay^{-1})^l$				
$R-$		y^m		$a^{-1}xy$	a^n	$(ay^{-1})^l$				
$R+$		y^m		$a^{-1}xy$	a^n	$(ay^{-1})^l$	$x^{-1}ay^{-1}$			
$R-$		y^m			a^n	$(ay^{-1})^l$	$x^{-1}ay^{-1}$			
$X-$		y^m			a^n	$(ay^{-1})^l$				
$X+$		y^m			a^n	$(ay^{-1})^l$	$b^{-1}y^{-1}$			
$R+$		y^m			a^n	$(ay^{-1})^l$	$b^{-1}y^{-1}$	b^m		
$R-$					a^n	$(ay^{-1})^l$	$b^{-1}y^{-1}$	b^m		
$R+$					a^n	$(ay^{-1})^l$	$b^{-1}y^{-1}$	b^m	$(ab)^l$	
$R-$					a^n		$b^{-1}y^{-1}$	b^m	$(ab)^l$	
$R+$					a^n		$b^{-1}y^{-1}$	b^m	$(ab)^l$	$y^{-1}b^{-1}$
$R-$					a^n			b^m	$(ab)^l$	$y^{-1}b^{-1}$
$X-$					a^n			b^m	$(ab)^l$	

Fig. 8

This proves that $D(l,m,n)$ and $D(n,m,l)$ are isomorphic, though the method is rather long and tedious. A substantial simplication is achieved if we work in terms of relations rather than relators. Thus, starting from

$$D(l,m,n) = <x,y \mid x^l = y^m = (xy)^n = e> \ ,$$

the above computation may be informally paraphrased as follows.

Introduce the generator $a = xy$, so that

$$a^n = e, \ \ x = ay^{-1}, \ \ (ay^{-1})^l = e \ ,$$

and the presentation reduces to

$$<a,y \mid a^n = (ay^{-1})^l \ y^m = e> \ .$$

Replacing y^{-1} by b and inverting the last relation we obtain

$$<a,b \mid a^n = (ab)^l = b^m = e> \ .$$

With a reasonable amount of care, no damage is done by this loss of precision. A rather more striking example, worked out in paraphrase only, is as follows.

Example 6. Consider the group

$$T = <x,y,z \mid x = yzy^{-1}, \ y = zxz^{-1}, \ z = xyx^{-1}> \ .$$

Using the last relation to eliminate z, we see that the first two relations become equivalent and we have the group

$$B_3 = <x,y \mid xyx = yxy> \ .$$

Putting $xy = a$, we have

$$y = x^{-1}a, \ \ ax = x^{-1}a^2 \ .$$

Finally, letting $x = a^{-1}b$, we obtain the elegant alternative presentation

$$<a,b \mid a^3 = b^2>$$

of T.

Our final result in this section complements Proposition 5 and is surprisingly easy to prove.

Proposition 6. Given two finite presentations of the same group, one can be obtained from the other by a finite sequence of Tietze transformations.

Proof. Given two such presentations

$$<X \mid R(X) = e>, \quad <Y \mid S(Y) = e> \, ,$$

suppose that

$$X = X(Y), \quad Y = Y(X) \tag{4}$$

are two systems of equations expressing the generators X in terms of the generators Y, and vice versa. We now apply Tietze transformations en bloc to the first presentation in accordance with the scheme in Fig. 9.

$$
\begin{array}{llll}
X+: & X,Y & R(X) = e, \ Y = Y(X), \\
R+: & X,Y & R(X) = e, \ Y = Y(X), \ X = X(Y) \\
R+: & X,Y & R(X) = e, \ Y = Y(X), \ X = X(Y), \ R(X(Y)) = e, \\
R-: & X,Y & Y = Y(X), \ X = X(Y), \ R(X(Y)) = e, \\
R+: & X,Y & Y = Y(X), \ X = X(Y), \ R(X(Y)) = e, \ Y = Y(X(Y)) \qquad (5) \\
R-: & X,Y & X = X(Y), \ R(X(Y)) = e, \ Y = Y(X(Y)) \\
X-: & Y & R(X(Y)) = e, \ Y = Y(X(Y)) \\
R+: & Y & R(X(Y)) = e, \ Y = Y(X(Y)), \ S(Y) = e \\
R-: & Y & S(Y) = e
\end{array}
$$

Fig. 9

Remark 5. In the course of this proof, we have added $|Y|$ generators and removed $|X|$, and $R+, R-$ have been used

$$|X| + |R| + |Y| + |S|, \ 2(|R| + |Y|)$$

times, respectively.

6. In contrast to this theorem, there is no general algorithm for deciding whether two given finite presentations yield isomorphic groups. (The apparent paradox is explained by the fact that, in proving the theorem, we made implicit use of an isomorphism to write down the equations (4)). This is known as the Isomorphism Problem, and is the hardest of three problems identified by M. Dehn. The other two are called the Word Problem and the Conjugacy Problem and are also undecidable in general.

7. We warn the reader against a popular error. Given a presentation $G = <X \mid R(X) = e>$
and new generators Y such that $X = X(Y)$, it is in general false that G is presented in terms
of Y by $<Y \mid R(X(Y)) = e>$. The correct presentation on the new generators is that in line
(5) of Fig. 9, namely,

$$G = <Y \mid R(X(Y)) = e, \ Y = Y(X(Y))> \ .$$

We conclude with an example of this process of changing generators that will be useful
later on.

Example 7. Let $m,n \in \mathbb{N}$ be coprime, so that, by Euclid's theorem

$$um + vn = 1 \ , \tag{6}$$

for some $u,v \in \mathbb{Z}$. Starting from

$$Z_{mn} = <a \mid a^{mn} = e> \ ,$$

we obtain a presentation on new generators $x = a^m$, $y = a^n$. Because of (6), $x^u y^v = a$, and
equations (4) read

$$a = x^u y^v \ \left| \begin{array}{l} x = a^m \\ y = a^n \ , \end{array} \right.$$

and the relations in (5) are

$$(x^u y^v)^{mn} = e, \ x = (x^u y^v)^m, \ y = (x^u y^v)^n \ .$$

Thus

$$<x,y \mid (x^u y^v)^{mn} = e, \ x = (x^u y^v)^m, \ y = (x^u y^v)^n> \tag{7}$$

is the presentation Z_{mn} on the new generators. New relations can be deduced as follows.
The last two relations assert that x and y are powers of a common element, whence

$$[x,y] = e \tag{8}$$

holds in the group presented by (7), as do

$$x^n = e, \ y^m = e \ , \tag{9}$$

by substituting the second and third relations, respectively, into the first. Conversely, the
relations in (7) are consequences of (8) and (9), and three $R+$'s followed by three $R-$'s

transform (7) into

$$<x,y \mid x^n = y^m = [x,y] = e> .$$

It now follows from Proposition 4 that

$$Z_{mn} \cong Z_m \times Z_n , \quad \text{when } (m,n) = 1 . \tag{10}$$

5.   van Kampen diagrams

At one level, van Kampen diagrams may be viewed as a pictorial method of illustrating the fact that a given word $w \in F(X)$ is a relation in a group $G = <X \mid R>$, i.e., that w is a product of conjugates of the words in $R^{\pm}$. At a more rigorous level, they (or rather their duals, which are called *star graphs*) form the basis of one of the most powerful methods of Combinatorial Group Theory. We adopt the former viewpoint here, and mention the latter only briefly in passing (see also the Guide to the Literature).

Informally then, a *van Kampen diagram* for the presentation $G = <X \mid R>$ is a finite connected planar graph $\Gamma \subseteq \mathbb{R}^2$ whose edges are directed and labelled by elements of X in such a way that every face of X (bounded component of $\mathbb{R}^2 \backslash X$) is a disc whose boundary label (for some starting point and orientation) belongs to R. It is then fairly clear (and not hard to prove) that the boundary label of Γ itself is a relation in G. van Kampen diagrams may thus be used to illustrate the deduction of new relations from old.

Before giving examples of this, we note that the above remark has a converse. Namely, *every* relator in G is the boundary label of some van Kampen diagram Γ for G. Moreover, it may be assumed that Γ is *reduced*, in the sense that no non-trivial circuit in Γ carries a label that reduces in $F(X)$ to the empty word. Now the fact that a graph Γ embeds in the plane imposes severe restrictions on its structure (for example, on the Euler characteristic: see §2.6). These may be used to argue from local properties of Γ (corresponding to combinatorial conditions in the set R) to properties of its boundary (corresponding to group-theoretical properties of G).

Another way to construct a van Kampen diagram for a presentation $G = <X \mid R \mid>$ is as follows. We think of each relator as the boundary label of a 2-cell. Collections of these may then be pasted together along edges with the same label and orientation.

Example 8. The quaternion group of order 8 is usually presented in the form

$$Q_4 = <x,y \mid x^4 = e, \; x^2 = y^2, \; y^{-1}xy = x^{-1}> .$$

Convert to relators by setting

$$r = x^4, \ s = x^2 y^{-2}, \ t = y^{-1} xyx \ ,$$

and consider the van Kampen diagram depicted in Fig. 10. In the top left face, the

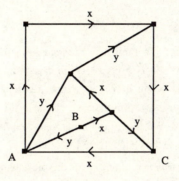

Fig. 10

boundary label read clockwise from the point A is $s = x^2 y^{-2}$, as is that of the upper face containing the point B, read counter-clockwise from B. In the lower face containing B, the boundary label read clockwise from A is $t = y^{-1} xyx$, as is that of the rightmost face read clockwise from C. Since the label on the boundary of the diagram itself, read clockwise from A, is just $r = x^4$, this proves that the first relation in the given presentation of Q_4 is superfluous.

Example 9. The group

$$F(2,4) = <a,b,c,d \mid ab = c, \ bc = d, \ cd = a, \ da = b>$$

(see §7.3) is clearly generated by a and b, as $c = ab$ and $d = bc = bab$. The following diagrams show that it is actually cyclic (Fig. 11: $a = b^{-3}$) of order 1 or 5 (Fig. 12: $b^5 = e$). Here the faces are labelled by the defining relators

$$r_1 = abc^{-1}, \ r_2 = bcd^{-1}, \ r_3 = cda^{-1}, \ r_4 = dab^{-1} \ .$$

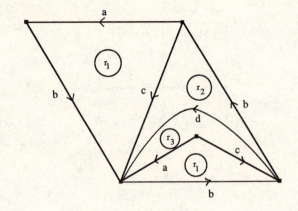

Fig. 11

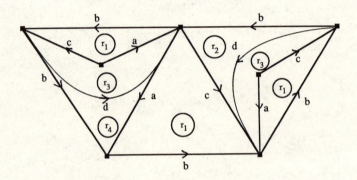

Fig. 12

Example 10. If a,b,c,d are generators of a group such that a and b each commute with each of c and d, then the fact that ab commutes with cd is expressed by the diagram in Fig. 13.

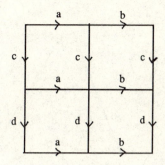

Fig. 13

Example 11. Finally, we prove (in Fig. 14) the non-obvious fact that

$$\underset{r}{x^2yxy^3} = e = \underset{s}{y^2xyx^3} \;=> x^7 = e$$

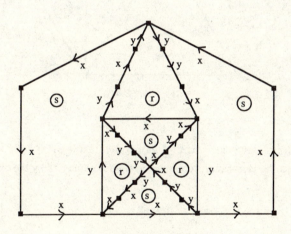

Fig. 14

Exercises

1. Give an alternative proof of Proposition 1 by showing that if G is any group, then G has the presentation $<X \mid M>$, where $X = G$ and

$$M = \{xy = (xy)m \mid x, y \in G\}$$

is the multiplication table of G (where $(x,y)m$ stands in the (x,y)-place).

2. Let $G = <X \mid R>$ and G' be the commutator subgroup of G. Prove that $G_{ab} = G/G'$ has the presentation

$$<X \mid R, \ [X,X]> \ ,$$

where $[X,X] = <[x,y] \mid x, y \in X, x \neq y\}$. Use Proposition 3 to confirm the result of Exercise 1.7.

3. For $l,m,n \in \mathbb{Z}$, prove that

$$D(l,m,n) \cong D(m,l,n) \cong D(-l,m,n) \ ,$$

and deduce that $D(l,m,n)$ is independent of the orders and signs of l,m,n.

4. Identify $D(l,m,n)$ in the following cases:

(a) one of l,m,n is 1, (b) two of l,m,n are 2.

5. Let $D_\infty = <a,b \mid a^2, b^2>$. Idenitfy D_∞/D'_∞. Prove that D'_∞ is a) cyclic, and b) infinite. [Hint: For a), consider conjugates of $[a,b]$, and for b), consider homomorphisms from D_∞ onto the groups $D(2,2,n)$].

6. Show that the groups

$$<a,b \mid a^b b^a = (b^{-1}a^2)^2 = e>, \quad <x,y \mid x^2 = y^3 = e>$$

are isomorphic.

7. Do you recognise the group

$$<a,b,c \mid a^3 = b^2 = c^2 = (ab)^2 = (bc)^2 = [a,c] = e> \ ?$$

What is the order of the element abc?

8.   Prove the converse of the result in (10) of Example 7:

$$(m,n) \neq 1 \Rightarrow Z_{mn} \not\cong Z_m \times Z_n \ .$$

9.   Prove that the groups

$$<x,y \mid xyx = y, \ yxy = x>, \ \ <a,b \mid a^2 = b^2, \ a^{-1}ba = b^{-1}>$$

are isomorphic.

10. Show that the groups

$$<a,b,c,d \mid ab = c, \ bc = d, \ cd = a, \ da = b>$$

$$<a,b,c,d,f,g \mid ab = d, \ bc = f, \ cd = g, \ df = a, \ fg = b, \ ga = c>$$

are cyclic and find their orders.

11. Consider the group $F(2,6)$ given by

$$<a,b,c,d,f,g \mid ab = c, \ bc = d, \ cd = f, \ df = g, \ fg = a, \ ga = b> \ .$$

Prove that there is a homomorphism $\chi : F(2,6) \to \text{Sym}(\mathbb{Z})$ such that

$$n(a\chi) = n + 1, \ n(b\chi) = -n, \ n \in \mathbb{Z} \ ,$$

and deduce that $F(2,6)$ is an infinite group.

12. Prove that the group $F(2,5)$ given by

$$<a,b,c,d,f \mid ab = c, \ bc = d, \ cd = f, \ df = a, \ fa = b>$$

is a finite cyclic group.

13. Check the *Witt Identity*

$$[[x,y],z^x] \ [[z,x],y^z] \ [[y,z],x^y] = e \ ,$$

and use it to prove that the group

$$<x,y,z \mid [x,y] = z, \ [y,z] = x, \ [z,x] = y>$$

is trivial.

14. Let x and y be members of a group such that $x^{-1}y^n x = y^m$, for some $m, n \in \mathbb{Z}$. Prove that, for any $k \in \mathbb{N}$, $x^{-1}y^{n^k} x = y^{m^k}$. Deduce that the group

$$<x, y \mid y^{-1}x^n y = x^{n+1}, x^{-1}y^n x = y^{n+1}>$$

is trivial for any $n \in \mathbb{Z}$.

15. Given $G = <X \mid R>$, prove that

    a) if G is finitely generated, then there is a finite subset $X' \subseteq X$ such that $G = <X'>$,

    b) if G is in addition finitely presented, then there is in addition, a finite subset of R' of R such that $G = <X' \mid R'>$.

16. Solve the word problem for finite groups according to the following scheme:

    a) let $G = <X \mid R>$ be a finite presentation of a finite group G, and let $w \in F(X)$;

    b) enumerate the elements of $\bar{R} \triangleleft F(X)$ as products of conjugates of $R^{\pm}$;

    c) enumerate the homomorphisms: $G \to S_n$, $n \in \mathbb{N}$ as mappings: $X \to S_n$ that vanish on R, and

    d) by examining alternate entries in the lists b) and c), decide in a finite number of steps whether $w = e$ in G or not.

17. Let G be the group

$$<x, y \mid x^2 y = y^2 x, \ x^8 = e> \ .$$

Prove that $y^8 = e$ in G and draw the associated van Kampen diagram.

18. Draw a van Kampen diagram for the group

$$<x, y \mid x^3 = y^2, \ x^3 = e, \ y^{-1}xy = x^2> \ ,$$

which shows that the second relation is superfluous.

19. Draw a van Kampen diagram which shows that the group

$$<a, b \mid abab^2 = e = baba^2>$$

is abelian.

CHAPTER 5

SOME POPULAR GROUPS

The problem to be addressed here is that of finding a presentation P for a group G specified in some other way. The general method of attack consists of three steps:

1. find a suitable set X of generators for G,

2. in terms of X, write down some relations $R = e$ that hold in G and hope that they are enough to define G,

3. letting $P = <X \mid R>$, show that the epimorphism $\theta'' : P \twoheadrightarrow G$ of Proposition 4.2 is one-to-one.

The last step is usually the hardest, and the method of its execution depends on the form of the original G. For example, it may be possible to bound $|G|$ below by using concrete information, and to bound $|P|$ above, by manipulation within P. If these bounds are equal and finite, then θ'' is an isomorphism. More generally, one may find a "normal form" for the elements of P, that is, compile a list L of words in $X^{\pm}$ such that every element of P is equal to onne member of L. The examples which follow illustrate the various techniques involved.

1. The quaternions

We begin with a problem that arises early in a first course on group theory, that of classifying groups of order eight, which may be paraphrased as follows.

Let G be a group of order eight and $x \in G$ an element of maximal order m. Then $m = 2$, 4 or 8 by Lagrange's theorem. In the first case, G is elementary abelian $(Z_2 \times Z_2 \times Z_2)$ and in the last is cyclic (Z_8). There remains the case when $H := <x>$ has order four, so that $H \triangleleft G$. Letting $y \in G \setminus H$, it follows that $y^2, y^{-1}xy \in H$. Simple arguments then show that

$$y^{-1}xy = x \text{ or } x^{-1}, \text{ and } y^2 = e \text{ or } x^2 .$$

When $y^{-1}xy = x$, G is abelian, and isomorphic to $Z_4 \times Z_2$ in both cases. If $y^{-1}xy = x^{-1}$ and $y^2 = e$, then $G \cong D_4$ (see Exercises 2.4 and 4.4). We are left with the following question: is there a group G of order 8 containing an element x of order 4 and an element $y \in G \setminus \{x\}$ such that $y^{-1}xy = x^{-1}$ and $y^2 = x^2$? There is at most one such group, since the given information implies that its elements must be

$$x^i, x^iy, \quad 0 \le i \le 3 ,$$

with multiplication table

	x^j	x^jy
x^i	x^{i+j}	$x^{i+j}y$
x^iy	$x^{i-j}y$	x^{i-j+2}

where powers of x are reduced modulo 4. To show that this does indeed define a group, one may either check the associative law directly (tedious) or make appeal to a concrete group (as follows).

Let $\mathbb{H}$ denote the set of matrices of the form

$$A = \begin{pmatrix} z & w \\ -\bar{w} & \bar{z} \end{pmatrix},$$

where $w, z \in \mathbb{C}$. It is easy to show that if $A, B \in \mathbb{H}$, then so do $A + B$, $-A$, AB, and A^{-1} (when $A \ne 0$): note that if $z = a + ib$ and $w = c + id$ are not both zero, then neither is

$$\det A = z\bar{z} + w\bar{w} = a^2 + b^2 + c^2 + d^2 .$$

$\mathbb{H}$ then forms a division algebra over $\mathbb{R}$ called the *quaternions*. Putting each of a,b,c,d equal to 1 in turn, and the others equal to 0, we obtain the basis

$$\begin{pmatrix} 1 & 0 \\ 0 & 1 \end{pmatrix}, \ X = \begin{pmatrix} i & 0 \\ 0 & -i \end{pmatrix}, \ Y = \begin{pmatrix} 0 & 1 \\ -1 & 0 \end{pmatrix}, \ \begin{pmatrix} 0 & i \\ i & 0 \end{pmatrix},$$

These matrices satisfy the familiar equations of Hamilton, and thus, together with their negatives, comprise a group Q_4 of order 8. It is easy to check the relations $X^4 = I_2$, $X^2 = Y^2$, $Y^{-1}XY = X^{-1}$, and so Q_4 is the fifth group of order 8:

$$Q_4 = \langle x, y \mid x^4 = e, \ x^2 = y^2, \ y^{-1}xy = x^{-1} \rangle .$$

2.   The Heisenberg group

This is the group H of matrices of the form

$$\begin{pmatrix} 1 & r & s \\ 0 & 1 & t \\ 0 & 0 & 1 \end{pmatrix}, \; r,s,t \in \mathbb{Z} \; .$$

Putting

$$A = \begin{pmatrix} 1 & 1 & 0 \\ 0 & 1 & 0 \\ 0 & 0 & 1 \end{pmatrix}, \; B = \begin{pmatrix} 1 & 0 & 0 \\ 0 & 1 & 1 \\ 0 & 0 & 1 \end{pmatrix}, \; C = \begin{pmatrix} 1 & 0 & 1 \\ 0 & 1 & 0 \\ 0 & 0 & 1 \end{pmatrix},$$

it is easy to check that

$$A^k = \begin{pmatrix} 1 & k & 0 \\ 0 & 1 & 0 \\ 0 & 0 & 1 \end{pmatrix}, \; B^l = \begin{pmatrix} 1 & 0 & 0 \\ 0 & 1 & l \\ 0 & 0 & 1 \end{pmatrix}, \; C^m = \begin{pmatrix} 1 & 0 & m \\ 0 & 1 & 0 \\ 0 & 0 & 1 \end{pmatrix},$$

for all $k,l,m \in \mathbb{Z}$. Each element of H is then equal to one of the matrices

$$A^k B^l C^m = \begin{pmatrix} 1 & k & m+kl \\ 0 & 1 & l \\ 0 & 0 & 1 \end{pmatrix}. \tag{1}$$

In view of the relations

$$[A,B] = C, \; [C,A] = [C,B] = I_3 \; ,$$

we propose the presentation

$$P = <a,b,c \mid [a,b] = c, \; [c,a] = [c,b] = e> \tag{2}$$

for H.

What has been said so far may be summarised by saying that the map $\theta'' : P \to H$ sending a,b,c to A,B,C is an epimorphism. It remains to prove that θ'' is one-to-one, and this is carried out using a normal form as follows. Consider the set L of words

$$u = a^k b^l c^m, \; k,l,m \in \mathbb{Z} \; ,$$

in P, and the effect of multiplying u on the right by each of $c^{\pm 1}, b^{\pm 1}, a^{\pm 1}$.

First, it is obvious that

$$uc^{\pm 1} = a^k b^l c^{m \pm 1} \in L .$$

Next, since c commutes with b in P,

$$ub^{\pm 1} = a^k b^{l \pm 1} c^m \in L .$$

Finally, the first relation in (2) can be written in the form

$$a^{-1} ba = bc^{-1} .$$

Then for all $l \in \mathbb{Z}$,

$$a^{-1} b^l a = (bc^{-1})^l = b^l c^{-l} , \qquad (3)$$

since b commutes with c. Similarly

$$ab^l a^{-1} = b^l c^l . \qquad (4)$$

Thus,

$$ua^{\pm 1} = a^k b^l c^m a^{\pm 1}$$

$$= a^k b^l a^{\pm 1} c^m, \text{ since } [a,c] = e ,$$

$$= a^k a^{\pm 1} b^l c^{\pm l} c^m, \text{ by (3) and (4) },$$

$$= a^{k \pm 1} b^l c^{m \pm l} \in U .$$

We have thus shown that $Lx \leq L$ for all $x \in \{a,b,c\}^{\pm}$. If w is any word in $\{a,b,c\}^{\pm}$, it follows by induction on $l(w)$ that $Lw \subseteq L$. Since $e \in L$ (take $k = l = m = 0$), it follows that

$$w = ew \in Lw \subseteq L ,$$

which proves that every element of P is equal to some member of L. On the other hand, it is clear from (1) that distinct members of L are carried by θ'' into distinct matrices in H. It follows that θ'' is an isomorphism and H has the presentation given in (2).

3. Symmetric groups

As usual, let

$$S_n = \text{Sym} (\{1,2,...,n\})$$

denote the symmetric group of degree $n \in \mathbb{N}$. To obtain a presentation for S_n, we proceed in three steps, as suggested above.

Step 1. Suitable generators for S_n are obtained as follows. Recall that every permutation can be written as a product of (disjoint) cycles, so the cycles generate S_n. Since any cycle is a product of transposition:

$$(a_1 a_2 ... a_l) = (a_1 a_2)(a_1 a_3) ... (a_1 a_l) ,$$

the transpositions generate S_n. But if (ij) is any transposition, where $1 \leq i < j \leq n$ without loss of generality, then

$$(ij) = (i\ i+1)(i+1\ i+2)...(j-2\ j-1)(j-1j)(j-2\ j-1)...(i+1\ i+2)(i\ i+1).$$

It follows that S_n is generated by "adjacent transpositions"

$$(i\ i+1), \quad 1 \leq i \leq n-1 .$$

Step 2. These generators clearly satisfy the relations

$$(i\ i+1)^2 = e, \quad 1 \leq i \leq n-1 ,$$

$$((i\ i+1)(i+1\ i+2))^3 = e, \quad 1 \leq i \leq n-2 ,$$

$$(i\ i+1)(j\ j+1) = (j\ j+1)(i\ i+1), \quad 1 \leq i, j \leq n-1, \ |i-j| \geq 2 .$$

Thus prompted, we put

$$G_n = <x_1,...,x_{n-1} \mid R,S,T> , \tag{5}$$

where

$$R = \{x_i^2 \mid 1 \leq i \leq n-1\} ,$$

$$S = \{(x_i x_{i+1})^3 \mid 1 \leq i \leq n-2\} ,$$

$$T = \{[x_i,x_j] \mid 1 \leq i < j-1 < n-1\} .$$

There is thus an epimorphism $\theta'' : G_n \twoheadrightarrow S_n$ sending x_i to $(i\ i+1)$, $1 \leq i \leq n-1$, and it remains to prove that θ'' is one-to-one.

Step 3. Since $|S_n| \geq n!$, it is sufficient to prove that $|G_n| \leq n!$, and this is done by induction on n. When $n = 1$, $G_n = <\mid>$ is the trivial group, of order $1 \leq 1!$ We thus assume that $n \geq 2$ and that $G_{n-1} \leq (n-1)!$ Let H be the subgroup of G_n generated by $x_1,...,x_{n-2}$, and define

$$y_0 = e, \quad y_i = x_{n-1}...x_{n-i}, \quad 1 \le i \le n-1 .$$

Consider the subset

$$A = \{hy_i \,|\, h \in H, \ 0 \le i \le n-1\} ,$$

of G_n; we shall prove that $A = G_n$. The hard bit (and the crux of the whole proof) is to show that

$$Ax_j \subseteq A, \quad 1 \le j \le n-1 ,$$

so we consider the product $hy_i x_j$ in six possible cases.

(i) $i = 0, \ j < n-1 : hy_i x_j = hx_j \in Hy_0 \subseteq A$, as $j < n-1$.

(ii) $i = 0, \ j = n-1 : hy_i x_j = hex_{n-1} = hy_1 \in Hy_1 \subseteq A$.

(iii) $i > 0, \ j > n-i : hy_i x_j = h\,(x_{n-1} \,...\, x_j x_{j-1} \,...\, x_{n-i})x_j$

$$= hx_{n-1} \,...\, x_j x_{j-1} x_j \,...\, x_{n-i}, \text{ by } T ,$$

$$= hx_{n-1} \,...\, x_{j-1} x_j x_{j-1} \,...\, x_{n-i}, \text{ by } R \text{ and } S ,$$

$$= (hx_{j-1})x_{n-1} \,...\, x_{n-i}, \text{ by } T ,$$

$$\in Hy_i \subseteq A .$$

(iv) $i > 0, \ j = n-i : hy_i x_j = hx_{n-1} \,...\, x_{n-i} x_{n-i}$

$$= hx_{n-1} \,...\, x_{n-i+1}, \text{ by } R ,$$

$$= hy_{i-1} \in Hy_{i-1} \subseteq A .$$

(v) $i > 0, \ j = n-i-1 : hy_i x_j = hx_{n-1} \,...\, x_{n-i} x_{n-(i+1)}$

$$= hy_{i+1} \in Hy_{i+1} \subseteq A .$$

(vi) $i > 0, \ j < n-i-1 : hy_i x_j = (hx_j)y_i, \text{ by } T ,$

$$\in Hy_i \subseteq A .$$

This shows that $hy_i x_j \in A$, for all $h \in H, 0 \le i \le n-1$ and $1 \le j \le n-1$, that is, that $Ax_j \subseteq A$, for all j. Because of R, we have

$$Ax_j^{-1} = Ax_j \subseteq A, \quad 1 \le j \le n-1 ,$$

and so, just at in Section 2 above (with A in place of L), it follows that every element of G_n is equal to a member of A. Since the relations in G_n that involve $x_1,...,x_{n-2}$ are precisely those in G_{n-1}, the map

$$\left.\begin{aligned} \phi : G_{n-1} &\to G_n \\ x_i &\mapsto x_i \end{aligned}\right\} \quad 1 \le i \le n-2$$

defines a homomorphism. Since $\operatorname{Im} \phi = H$, by definition, and $|G_{n-1}| \le (n-1)!$ by hypothesis, we have

$$|G_n| \le |A| \le n |H| \le n! \ ,$$

as required.

4.   Semi-direct products

There are three ingredients in this construction: a group G, a group A, and a homomorphism $\alpha : G \to \operatorname{Aut} A$. The map α determines an *action* of G on A given by

$$a^x := a(x\alpha), \ a \in A, \ x \in G \ , \tag{6}$$

where the right-hand side is the image of $a \in A$ under the automorphism $x\alpha$ of A. Since α is a homomorphism,

$$(ab)^x = a^x b^x, \ a^{xy} = (a^x)^y, \ a^e = a \ , \tag{7}$$

for all $a,b \in A, x,y \in G$.

Now consider the Cartesian product $K = G \times A$, with the binary operation

$$(x,a)(y,b) = (xy, a^y b) \ . \tag{8}$$

The rules (7) guarantee that the group axioms hold for this operation: for example, the identity is (e,e) and

$$(x,a)^{-1} = (x^{-1}, (a^{-1})^{x^{-1}}) \ .$$

Definition 1. The group K just defined is called the *semi-direct product* of G and A (with respect to α), often written $A]G$.

The mappings

$$\left.\begin{aligned} A &\to K \\ a &\mapsto (e,a) \end{aligned}\right\}, \quad \left.\begin{aligned} G &\to K \\ x &\mapsto (x,e) \end{aligned}\right\}$$

are easily seen to be homomorphisms, and it is customary to identify A and G with their images in K. Then A is a normal subgroup of K, with complement G

$$A \lhd G, \ G \le K, \ A \cap G = E, \ GA = K \ . \tag{9}$$

The reason for the name "semi-direct product" is now clear: in the special case when $\text{Im}\,\alpha = E$, the action (6) is trivial, and the multiplication defined by (8) is just that in the direct product $G \times A$.

It is not difficult to write down a presentation for $A]G$ in terms of presentations for A and G and the map α. This is a "twisted" version of the presentation (4.3) for the direct product, and will be derived in a more general setting later. We content ourselves here by giving an example in the favourable special case when both A and G are finite cyclic groups. The resulting groups are *metacyclic*, that is, they each have a cyclic normal subgroup with cyclic factor group.

Let

$$A = Z_n = <y \,|\, y^n>, \quad G = Z_m = <x \,|\, x^m> \ , \tag{10}$$

so that Aut A is an abelian group of order $\phi(n)$, where ϕ is Euler's totient function. Now suppose that

$$y(x\alpha) = y^l, \ 0 \le l \le n - 1 \ . \tag{11}$$

Then there are two constraints on l. First, since $x\alpha$ is an automorphism, $|y^l|$ in Z_n is n, so that l must be coprime to $|y| = n$. Next, for $\alpha : Z_m \to \text{Aut}\,A$ to be a homomorphism, Proposition 4.3 requires that $(x\alpha)^m = 1_A$, that is,

$$l^m \equiv 1(\text{mod } n) \ . \tag{12}$$

Subject only to (12), which implies $(l,n) = 1$, (11) determines α completely, since

$$(y^r)(x^s\alpha) = ((y)(x\alpha)^s)^r = y^{rl^s} \ ,$$

for $0 \le r \le n - 1, \ 0 \le s \le m - 1$.

Now let $K = A]G$, with A and G as in (10), and with α given by (11), subject to (12). Regarding A and G as subgroups of K, the last equation in (9) asserts that

$$K = \{x^r y^s \,|\, 0 \le r \le m - 1, \ 0 \le s \le n - 1\} \ , \tag{13}$$

so that K is generated by x,y. Turning to relations, we clearly have

$$x^m = y^n = e \ ,$$

since these hold in G, A. Next, putting $(x, a) = (e, y) = y$ and $(y, b) = (x, e) = x$ in (8),

$$yx = xy^x, \ \text{i.e.} \ y^x = x^{-1}yx \ ,$$

which agrees with the usual notation for conjugates. From (6) and (11), on the other hand,

$$y^x = y(x\alpha) = y^l \ .$$

Combining these gives the relation $x^{-1}yx = y^l$, and we put

$$P = <x, y \mid x^m = y^n = e, \ x^{-1}yx = y^l> \ , \tag{14}$$

as a putative presentation for K.

From what has been said, the map $\theta'' : P \to K$ fixing x and y is an epimorphism. It remains to prove that θ'' is one-to-one, and for this, it is sufficient to prove that the right-hand side of (13), call it L, gives a normal form for elements of P (for then, $|P| = |K| = mn$). As in Sections 2, 3 above, we need only show that $Lx^{\pm 1}, Ly^{\pm 1} \subseteq L$, and since $x^{-1} = x^{m-1}, y^{-1} = y^{n-1}$ in P, this follows from the equations

$$x^r y^s y = x^r y^{s+1}, \ x^r y^s x = x^{r+1} y^{ls} \ ,$$

with powers reduced mod m and mod n, respectively. Thus, the semi-direct product $K = A]G$ given by (10), (11), (12) has the presentation (14).

5.   Groups of symmetries

Consider the lattice of integer points in $\mathbb{R}^2$:

$$L = \{(k, l) \in \mathbb{R}^2 \mid k, \ l \in \mathbb{Z}\} \ ,$$

and the group $G = \text{Sym } L$ of isometries of $\mathbb{R}^2$ that carry L onto itself.

One example of such an isometry is a translation t. If t carries the origin $(0,0)$ to the point (m, n), then it carries an arbitrary point (k, l) into $(k + m, \ k + n)$. t is thus determined by the pair (m, n), and if we let a, b denote the translations carryig $(0,0)$ to $(1,0)$, $(0,1)$, respectively, then $t = a^m b^n$. It t' maps $(0,0)$ to (k, l), then

$$(0,0)t't = (k, l)t = (k + m, \ l + n) \ .$$

In other words

$$a^k b^l a^m b^n = a^{k+m} b^{l+n} \ ,$$

and so the translations comprise a subgroup T of G. From what has been said, it is clear that

$$T = <a \mid > \times <b \mid > = <a,b \mid [a,b] = e> , \tag{15}$$

a free abelian group of rank 2.

Next consider isometries S that fix the origin 0, such as rotation y counter-clockwise about 0 through $\pi/2$, and reflection x in the line through 0 and $(0,1)$, that is, in the y-axis. In terms of matrices operating on the usual Cartesian coordinates (on the right), we have

$$x = \begin{pmatrix} -1 & 0 \\ 0 & 1 \end{pmatrix}, \ y = \begin{pmatrix} 0 & 1 \\ -1 & 0 \end{pmatrix}. \tag{16}$$

Now every element of S induces a unique symmetry of the square with vertices $(\pm1,0)$, $(0,\pm1)$, and so there can be at most eight such symmetries (since they map the four vertices among themselves and preserve adjacency). Thus, $S = <x,y>$ has order eight, and since the relations $y^4 = x^2 = e, x^{-1}yx = y^{-1}$ hold in S and define a group of order eight,

$$S = <x,y \mid x^2 = y^4 = e, \ x^{-1}yx = y^{-1}> , \tag{17}$$

a copy of the dihedral group D_4.

Now let $g \in G$ be arbitrary. If $0g = (m,n)$, and $t = a^m b^n \in T$, then gt^{-1} fixes 0 and thus belongs to S. Putting $gt^{-1} = s \in S$, we have $g = st$, whence $G = ST$. Since a translation that fixes any point must be the identity, it is clear that $S \cap T = E$, and G is beginning to look like a semi-direct product (see (9)). To prove that this is indeed the case, let $P = (m,n)$ be a typical point, and observe that, by (16),

$$Px^{-1}ax = (-m,n)ax = (-m + 1,n)x = (m - 1,n) = Pa^{-1} ,$$

$$Px^{-1}bx = (-m,n)bx = (-m,n + 1)x = (m,n + 1) = Pb ,$$

$$Py^{-1}ay = (n, -m)ay = (n + 1,-m)y = (m,n + 1) = Pb ,$$

$$Py^{-1}by = (n, -m)by = (n, -m + 1)y = (m - 1,n) = Pa^{-1} .$$

Hence,

$$a^x = a^{-1}, \ b^x = b, \ a^y = b, \ b^y = a^{-1} , \tag{18}$$

so that $T = <a,b>$ is normalized by $S = <x,y>$. Since $G = ST$, it follows that $T \lhd G$. Equations (18) show that the matrices (16) define a homomorphism

$$\alpha : S \to GL\,(2,\mathbb{Z}) = \text{Aut}\,T \ ,$$

whence $G = T]S$ is the semi-direct product of S and T with respect to α. An argument similar to that at the end of the previous section (to be justified in full later) shows that (15), (17), (18) can be combined to give

$$G = <x,y,a,b \mid [a,b] = x^2 = y^4 = e, \ y^x = y^{-1}, \ a^x = a^{-1}, \ b^x = b, \ a^y = b, \ b^y = a^{-1}> \ .$$

6.   Polynomials under substitution

Given a prime p and a positive integer n, consider polynomials

$$f(x) = \sum_{k=0}^{n} a_k x^n$$

of degree at most n over the field $\mathbb{Z}_p$ of p elements such that $a_0 = 0$ and $a_1 = 1$. It is not hard to show that, under the binary operation of composition of functions mod x^{n+1}, these form a group $G_n(p)$ of order p^{n-1}. These groups have many interesting propeties, but we confine ourselves here to finding a presentation for the group $G = G_5(2)$ of order 16. Thus, in the calculations that follow, coefficients are mod 2 and all powers of x higher than the fifth are zero.

First, let $a = x + x^2$. Then

$$a^2 = x + x^2 + (x + x^2)^2 = x + x^4 \ ,$$

$$a^3 = x + x^2 + (x + x^2)^4 = x + x^2 + x^4 \ ,$$

$$a^4 = (a^2)^2 = x + x^4 + (x + x^4)^4 = x \ ,$$

which is the identity in G. Thus, $A := <a>$ has order 4.

Next, let $b = x + x^3$. Then a similar calculation shows that

$$b^2 = x + x^5 \ ,$$

$$b^3 = x + x^3 + x^5 \ ,$$

$$b^4 = x \ .$$

Thus, $B := <b>$ has order 4. As $A \cap B = E$, AB must have 16 elements, so that $G = AB$. In this case, neither A nor B is normal in G, so that G is not their semi-direct product. One readily checks however that

$$ab = x + x^2 + x^3, \ ab^{-1} = x + x^2 + x^3 + x^5 \ ,$$

and then that

$$(ab)^2 = x = (ab^{-1})^2 \ .$$

It follows that G is a homomorphic image of the group

$$P = <a,b \,|\, a^4 = b^4 = (ab)^2 = (ab^{-1})^2 = e> \ . \tag{19}$$

To prove that P has order 16, we have only to show that the set

$$L = \{a^i b^j \,|\, 0 \le i, j \le 3\}$$

exhausts P. For this, it is sufficient (as in previous sections) to show that La, $Lb \subseteq L$. The second of these is obvious, and the first reduces to proving that ba, $b^2 a$, $b^3 a \in L$. The third relation of (19) gives

$$ba = a^{-1} b^{-1} = a^3 b^3 \in L \ .$$

The fourth relation of (19) gives

$$b^3 a = b^{-1} a = a^{-1} b = a^3 b \in L \ .$$

Finally,

$$abab = ab^{-1} ab^{-1} \Rightarrow bab = b^{-1} ab^{-1} \Rightarrow b^2 a = ab^2 \in L \ .$$

It follows that G has the presentation given in (19).

7. The rational numbers

Let $Q(+)$ denote the group of rational numbers under addition. The elements

$$x_n = 1/n! \ n \in \mathbb{N}$$

clearly generate $Q(+)$: if $a/b \in Q(+)$ with $a \in \mathbb{Z}$ and $b \in \mathbb{N}$, then

$$a/b = (a(b-1)!)x_b \ .$$

Since the x_n satisfy relations

$$nx_n = x_{n-1}, \ n \ge 2 \ , \tag{20}$$

we propose the presentation

$$P = \langle x_n, n \geq 1 \mid x_n^n = x_{n-1}, \ n \geq 2 \rangle \ . \tag{21}$$

Since any pair of generators of P commute (as one is a power of the other), P is an abelian group. For convenience, we write it in additive form

$$\langle x_n, n \geq 1 \mid n x_n = x_{n-1}, n \geq 2 \rangle \ . \tag{22}$$

Any additive word w in the x_n can thus be expressed in the form

$$w = \sum_{n=N}^{1} a_n x_n, \ N \in \mathbb{N}, \ a_n \in \mathbb{Z}, \ 1 \leq n \leq N \ . \tag{23}$$

This expression is said to be in *normal form* if

$$a_N \neq 0, \ 0 \leq a_n \leq n - 1 \text{ when } N \geq n \geq 2, a_1 \text{ arbitrary} \ . \tag{24}$$

The aim is now to reduce an arbitrary word w to normal form using the relations (20), and this is achieved by adjusting each a_n in turn starting on the left with a_N.

Thus, assume that (24) holds in w for all $n > k$, where $N \geq k \geq 2$, and put

$$a_k = kq + a'_k, q \in \mathbb{Z}, 0 \leq a'_k < k \ .$$

Then

$$a_k x_k = a'_k x_k + q \, (k x_k)$$

$$= a'_k x_k + q x_{k-1} \ ,$$

using (20). Then

$$w = a_N x_N + \dots + a_k x_k + a_{k-1} x_{k-1} + \dots + a_1 x_1$$

$$= a_N x_N + \dots + a'_k x_k + a'_{k-1} x_{k-1} + \dots + a_1 x_1 \ ,$$

where $a'_{k-1} = a_{k-1} + q$, and (24) now holds for all subscripts $\geq k$. The induction starts at $k = N$ with a vacuous assumption, and finishes at $k = 2$ with the required normal form.

It remains to prove that distinct normal forms have distinct images in $Q \, (+)$. To this end, assume for a contradiction that w, as in (23), and

$$w' = \sum_{n=M}^{1} b_n x_n$$

are distinct normal forms, but that their images in $Q(+)$ coincide, that is

$$\sum_{n=1}^{N} a_n/n! = \sum_{n=1}^{M} b_n/n!$$ (25)

Before proceeding, we need the following little lemma.

Lemma 1. Let $k, M \in \mathbb{N}$ with $k \leq M$, and for each integer n, $k + 1 \leq n \leq M$, let $a_n \in \mathbb{Z}$ satisfy $0 \leq a_n < n$. Then

$$\sum_{n=k+1}^{M} a_n/n! \leq \frac{1}{k!} - \frac{1}{M!}$$ (26)

Proof. For a fixed k, proceed by induction on $M \geq k$. When $M = k$ the result is trivial, so let $M > k$ and assume the result for $M - 1$. Then

$$\sum_{n=k+1}^{M} a_n/n! = \sum_{n=k+1}^{M-1} a_n/n! + a_M/M!$$

$$\leq \frac{1}{k!} - \frac{1}{(M-1)!} + \frac{M-1}{M!}$$

$$= \frac{1}{k!} - \frac{1}{M!} ,$$

as required.

Now suppose that w and w' first differ in the k-place, $k \geq 1$, and that $a_k > b_k$, without loss of generality. Then (25) becomes

$$a_k/k! + \sum_{n=k+1}^{N} a_n/n! = b_k/k! + \sum_{n=k+1}^{M} b_n/n! .$$

Then, using (26),

$$\frac{1}{k!} \leq (a_k - b_k)/k! \leq (a_k - b_k)/k! + \sum_{n=k+1}^{N} a_n/n!$$

$$= \sum_{n=k+1}^{M} b_n/n! \leq 1/k! - 1/M! < 1/k! ,$$

which is the required contradiction.

It follows that $Q(+)$ has the presentation (21).

Exercises

1. Let $w = e^{i\pi/n} \in \mathbb{C}$. Prove that the matrices

$$X = \begin{pmatrix} w & 0 \\ 0 & \overline{w} \end{pmatrix}, \quad Y = \begin{pmatrix} 0 & 1 \\ -1 & 0 \end{pmatrix}$$

generate a subgroup Q_{2n} of order $4n$ in $GL(2, \mathbb{C})$, with presentation

$$<x,y \mid x^n = y^2, \ x^{2n} = e, \ y^{-1}xy = x^{-1}> .$$

(Q_{2n} is called a *generalised quaternion group*).

2. Consider the group

$$Q = Hop \setminus \{0\} = \left\{ \begin{pmatrix} z & w \\ -\overline{w} & \overline{z} \end{pmatrix} \middle| z, w \in \mathbb{C}, \ \text{not both zero} \right\} .$$

Prove that

(i) $Z(Q)$ consists of the real scalar matrices, and

(ii) Q' consists of the matrices of determinant 1. Putting $\tilde{Q} = Q/\{\pm I_2\}$, deduce that

(iii) $\tilde{Q} = Z(\tilde{Q}) \times \tilde{Q}'$,

(iv) $\tilde{Q}'' = \tilde{Q}'$, and

(v) $Z(\tilde{Q}') = E$.

3. As an alternative to the method given in Section 2, put

$$P = <a,b \mid [[a,b], a] = [[a,b], b] = e> ,$$

and show that P' is an infinite cyclic group generated by $[a,b]$. Deduce that every element of P is uniquely expressible in the form $a^k b^l [a,b]^m$.

4. Prove that the group of matrices

$$\left\{ \begin{pmatrix} 1 & r & s \\ 0 & 1 & t \\ 0 & 0 & 1 \end{pmatrix} \in GL(3, \mathbb{Z}_n) \middle| r,s,t \in \mathbb{Z}_n \right\}$$

over the integers modulo n has the presentation

$$<a,b,c \,|\, a^n = b^n = c^n = e, \ c = [a,b], \ [c,a] = [c,b] = e> \ .$$

5.   Let $k,l,m \in \mathbb{N}$. Calculate the order of the group

$$<a,b,c \,|\, a^k = b^l = c^m = e, \ [a,b] = c, \ [c,a] = [c,b] = e> \ .$$

6.   Use Tietze transformations to convert the presentation (5) with $n = 4$ to a presentation on the generators $x = (12)$, $y = (234)$. Deduce that $S_4 \cong D\,(2,3,4)$.

7.   Verify the group axioms for the binary operation given by (8).

8.   Prove that Aut Z_n is an abelian group of order $\phi(n)$, $n \geq 1$, and that Aut $Z \cong Z_2$.

9.   What is the order of the gorup (14) when $l \in \mathbb{N}$ is arbitrary (that is, does not necessarily satisfy (12))?

10. Find presentations of $Z_n]Z$, $Z\,]Z_m$, $m,n \in \mathbb{N}$, for all possible actions.

11. Write down presentations for the five groups of order p^3, where p is prime.

12. Let L be the tesselation of the euclidean plane by equilateral triangles. Prove that Sym (T) is a semi-direct product $(Z \times Z)]D_6$, and obtain a presentation of this group.

13. Let G_4 be the group $\{x + ax^2 + bx^3 + cx^4 \,|\, a,b,c \in \mathbb{Z}\}$ under polynomial substitution molulo x^5.. Prove that G_4 is isomorphic to the Heisenberg group H.

14. Up to isomorphism, there are 14 groups of order 16. Of these, all but one are mentioned more or less explicitly in this chapter. Can you identify them? Prove that the group

$$<x,y,z \,|\, x^2 = y^2 = z^2 = e, \ xyz = yzx = zxy>$$

has order 16, thereby completing the list. [Hint: of the 14 groups, 5 are abelian. Of the other 9, 4 contain a cyclic subgroup H of order 8, and 3 of these have the form $H]Z_2$. Of the remaining 5, two are non-trivial direct products and one is metacyclic.]

15. Prove that every non-trivial finitely-generated subgroup of $Q\,(+)$ is infinite cyclic, but that $Q\,(+)$ itself is not.

CHAPTER 6

FINITELY-GENERATED ABELIAN GROUPS

1. Groups-made-abelian

Given any group G, recall that its derived group (or commutator subgroup) is the group G' generated by the set of all commutators $\{g^{-1}h^{-1}gh \mid g,h \in G\}$ of elements of G. It is clear that $G' \lhd G$ and that $G_{ab} := G/G'$ is abelian. Moreover, G' is the "smallest" subgroup of G with this property (so that G_{ab} is the "largest" abelian factor group of G) in the following precise sense:

$$\text{given } N \lhd G, \; G/N \text{ is abelian } \Leftrightarrow N \supseteq G' \; . \tag{1}$$

G_{ab} is often called the *abelianization* of G or "G-made-abelian", and is an invariant of G. (It is with regret that we use the word "abelianization", which must be among the most hideous in the English language).

In the context of presentations, two questions thus arise in a natural way. Given a presentation $P = <X \mid R>$ of a group G,

    (i)   how can we extract from P information about G_{ab}?

    (ii) how do we interpret the information thus extracted?

It turns out that, when X is finite, both questions can be answered in an entirely satisfactory way. The answer to the second question is a perfect example of a classification theorem: it is possible to draw up a list of $f.g.$ abelian groups with the property that *every* $f.g.$ abelian group is isomorphic to *onne* group in the list. This is the celebrated basis theorem (for $f.g.$ abelian groups). We prove it in full here for two reasons: (i) it is one of the most important theorems of undergraduate mathematics, and (ii) its proof contains a simple numerical algorithm that answers question (i), above. We are thus in the happy position of being able to kill two birds with one stone.

Let us fix the notation

$$X = \{x_1, x_2, ..., x_r\}, \; C = \{[x_i, x_j] \mid 1 \leq i < j \leq r\}, \; r \in \mathbb{N} \; , \tag{2}$$

where, by abuse of notation, C may be regarded as a subset of any group presented on generators X. In view of its fundamental importance here, we elevate Exercise 4.2 to the status of a proposition.

Proposition 1. If $G = <X \mid R>$, then $G_{ab} = <X \mid R,C>$.

Proof. By von Dyck's theorem (Proposition 4.2) on the adjunction of relators, it is sufficient to prove that G' coincides with the normal closure $\overline{C}$ of C in G. Since the generators of $<X \mid R,C> = G/\overline{C}$ all commute, this group is abelian, and so $G' \subseteq \overline{C}$ by (1). On the other hand, G' is a normal subgroup of G containing C, when $\overline{C} \subseteq G'$.

Example. Let G be the group given by the presentation

$$<x,y,z,t \mid (xyz)^6 = e, \ t^2 = (xz)^2, \ (xy^3z^2)^2 = e \ ,$$

$$(yt^2)^2 = x^2z^3, \ (xyz)^4(yt)^2 = e> \ .$$

Then G_{ab} has the presentation

$$<x,y,z,t \mid (xyz)^6 = e, \ t^2 = (xz)^2, \ (xy^3zt^2)^2 = e \ ,$$

$$(yt^2)^2 = x^2z^3, \ (xyz)^4(yt)^2 = e \ ,$$

$$[x,y] = [x,z] = [x,t] = [y,z] = [y,t] = [z,t] = e> \ .$$

In view of the commutators, we may collect powers of like elements in the first five relations and convert these into relators to obtain the following more convenient form:

$$G_{ab} = <z,y,z,t \mid x^6y^6z^6, \ x^2z^2t^{-2}, \ x^2y^6z^2t^4,$$

$$x^{-2}y^2z^{-3}t^4, \ x^4y^6z^4t^2 \ ,$$

$$[x,y],[x,z],[x,t],[y,z],[y,t],[z,t]> \ . \tag{3}$$

The power of each generator in each of the first five relations is called its *exponent-sum*.

2.   Free abelian groups

We begin by similarly upgrading Exercise 1.7.

Proposition 2. If $F = F(X)$ is free of rank r, then F/F' is:

(i)   given by the presentation $<X \mid C>$,

(ii)   isomorphic to the direct product of r copies of the infinite cyclic group,

(iii) free abelian of rank r.

Proof. Part (i) is just the case $R = \varnothing$ of Proposition 1. To prove that the direct product $Z^{\times r}$ or r copies of Z has the presentation $<X \mid C>$, we proceed by induction on r, noting that C is empty when $r = 1$. For $r > 1$, assume that

$$Z^{\times(r-1)} = <x_1,...,x_{r-1} \mid C_{r-1}>, \quad C_{r-1} = \{[x_i,x_j] \mid 1 \leq i < j \leq r-1\} \ ,$$

whereupon

$$Z^{\times r} = Z^{\times(r-1)} \times Z = Z^{\times(r-1)} \times <x_r \mid >, \quad \text{say} \ ,$$

$$= <X \mid C_{r-1}, [x_1,x_r],...,[x_{r-1},x_r]>, \quad \text{by Proposition 4.4} \ ,$$

and this is just $<X \mid C>$.

Finally, to prove freeness, we assume that F^{ab} is the (internal) direct product of infinite cyclic groups generated by $x_1,...,x_r$, in accordance with part (ii). Changing to additive notation, every element of F^{ab} is uniquely a Z-linear combination of $x_1,...,x_r$. Now let G be any additively-written abelian group and $\theta : X \to G$ any mapping, and define $\theta' : F^{ab} \to G$ as follows. If $x \in F^{ab}$, we write $x = \sum\limits_{i=1}^{r} k_i x_i (k_i \in Z, \ 1 \leq i \leq r)$, and define

$$x\theta' = \sum\limits_{i=1}^{r} k_i(x_i\theta) \ .$$

This clearly extends θ and is easily seen to be a homomorphism (cf. the proof of freeness of $F(X)$ in Theorem 1.1); it is unique since X generates F^{ab}.

Remark 1. From now on, we denote the free abelian group on X by $A = A(X)$ and continue to write it additively. That its elements are unique $\mathbb{Z}$-linear combinations of elements of X (as in the above proof) is just the abelian analogue of Proposition 1.3. Our next step towards the Basis Theorem consists of proving the analogues of Proposition 1.4 and Theorem 2.1 (Nielsen-Schreier).

Proposition 3. If X generates an abelian group G, then there is an epimorphism $\theta : A(X) \to G$ fixing X elementwise. Every abelian group is a homomorphic image of

some free abelian group.

Proof. To prove this, we can either give an abelian version of the proof of Proposition 1.4 or proceed as follows. If $G = <X \mid R>$ is abelian, then $G = G/G' = <X \mid R, C>$ and, shifting our weight on to the other foot as it were, we see (by von Dyck's theorem) that G is just the factor group of $A(X) = <X \mid C>$ by $\overline{R}$.

Proposition 4. (Dedekind). If $A = A(X)$ is free abelian of rank r and B is a subgroup of A, then B is free abelian of rank at most r.

Proof. We go by induction on r, noting that the case $r = 1$ is just a well known result of elementary group theory. Now let $r > 1$ and assume the result for $r - 1$. Define subgroups

$$H = <x_1,...,x_{r-1}>, \quad C = <x_r>$$

of A, so that H is free abelian of rank $r - 1$ and $A = H \times C$. By the inductive hypothesis, $B \cap H$ is free abelian - on $y_1,...,y_s$ say, with $s \le r - 1$. Furthermore,

$$\frac{B}{B \cap H} \cong \frac{B + H}{H} \le \frac{A}{H} \cong C \ .$$

so that $B/B \cap H$ is either trivial or infinite cyclic (by the case $r = 1$). In the first case, $B = B \cap H$ and we are done, so we assume that $B/B \cap H = <b + B \cap H>$ with $b \in B \setminus H$. We write $b = h + lx_r$, $h \in H$, $l \in \mathbb{N}$ and claim that B is free abelian on the set $Y = \{y_1,...,y_s,b\}$. To do this, we invoke the abelian analogue of Proposition 1.3 (see Remark 1 above), noting first that Y clearly generates B. To prove $\mathbb{Z}$-linear independence of the set Y, suppose that

$$\sum_{i=1}^{s} k_i y_i + kb = 0, \ k_i, \ k \in \mathbb{Z} \ . \tag{4}$$

Thus we have

$$klx_r = k(b - h) = - \sum_{i=1}^{s} k_i y_i - kh \in H \ ,$$

whence $klx_r = 0$, since $H \cap <x_r> = \{0\}$. Since $l \ne 0$, we must have $k = 0$, and (4) reduces to $\sum_{i=1}^{s} k_i y_i = 0$, and since the y_i are free generators of $B \cap H$, each $k_i = 0$. Thus every element of B is uniquely a $\mathbb{Z}$-linear combination of the elements of Y, which proves our claim and hence the result.

Example. Referring to the above example, the group G_{ab} in (3) is the factor group A/B, where

$$A = A(x,y,z,t) \text{ is a free abelian group of rank } 4 ,$$

and $B = <6x + 6y + 6z, \ 2x + 2z - 2t, \ 2x + 6y + 2z + 4t,$ (5)

$$-2x + 2y - 3z + 4t, \ 4x + 6y + 4z + 2t> ,$$

is additive notation. Since the rank of B is at most four, these five elements do not generate B freely. We shall see later that B in fact has rank 3.

3.   Change of generators

From the results of the previous section, we now know that every $f.g.$ abelian group is isomorphic to a group of the form $A(X)/B$, where $A(X)$ is the free abelian group on $X = \{x_1,...,x_r\}$, and B is freely generated by a set $Y = \{y_1,...,y_s\}$ of $\mathbb{Z}$-linear combinations of elements of X. Noting that $B = <Y> = \overline{Y}$ (since $A(X)$ is abelian), and writing the elements of $Y = Y(X)$ multiplicatively (as words in $X^{\pm}$), we obtain the presentation

$$A(X)/B = <X \mid Y(X),C> .$$ (6)

Note for future reference this is of exactly the same type as the presentation for G_{ab} given in Proposition 1. Our aim is now to examine the effect of changing the free generators X of A and Y of B on the defining relators $Y(X)$, in order to make these as simple as possible.

Forgetting about Y and B for the moment, let $U = \{u_1,u_2,...,u_n\}$ be another set of *free* generators for A. Corresponding to the relations $X = X(U)$, $U = U(X)$ (see (4.4) in the proof of Proposition 4.6), we have systems of equations

$$x_i = \sum_{j=1}^{n} p_{ij}u_j, \ 1 \leq i \leq r ,$$ (7)

$$u_j = \sum_{k=1}^{r} q_{kj}x_k, \ 1 \leq j \leq n ,$$ (8)

where the p_{ij}, q_{jk} are integers. Substituting (8) in (7), uniqueness of expression yields that

$$\sum_{j=1}^{r} p_{ij}q_{jk} = \delta_{ij}, \ 1 \leq i,k \leq r .$$

In other words, the integer matrices $P = (p_{ij})$, $Q = (q_{jk})$ have the property that $PQ = I_r$. Similarly, the freeness of the generators U guarantees that $QP = I_n$. This proves that $r = n$ and so $Q = P^{-1}$. Conversely, any transformation of the type (8) with $Q = (q_{ik})$ invertible

over $\mathbb{Z}$ will yield a new set of free generators of A.

Now let B be an arbitrary subgroup of A, so that by Dedekind's theorem, B is free abelian of rank s, with $s \leq r$. Letting $Y = \{y_1,...,y_s\}$ be a set of free generators of B, we have equations

$$y_k = \sum_{i=1}^{r} m_{ki} x_i, \quad 1 \leq k \leq s , \qquad (9)$$

and B is determined by the $s \times r$ matrix $M = (m_{ki})$, with respect to the sets Y,X of free generators. The effect on M of changing to generators U of A is found by substituting (7) into (9), which yields the matrix $MP = MQ^{-1}$. On the other hand, if Y is changed to another set V of free generators of B in accordance with an invertible $s \times s$ matrix T, then B is determined with respect to V,X by TM, and with respect to V,U by TMQ^{-1}. We summarise these remarks in the form of a proposition.

Proposition 5. The subgroup $B = <Y>$ of $A(X)$ (and thus also of $A(X)/B$) is determined by the $s \times r$ coefficient matrix $M = (m_{ki})$ of (9). Changing the free generators X and Y corresponds to post- and pre-multiplication of M, respectively, by invertible matrices over $\mathbb{Z}$. Conversely, if T and Q^{-1} are unimodular, the coefficient matrix TMQ^{-1} determines the same subgroup B of $A(X)$ as does M (and thus the same factor group $A(X)/B$).

Example. In the case of our on-going example (see (5) above), the coefficient matrix is given by

$$M = \begin{pmatrix} 6 & 6 & 6 & 0 \\ 2 & 0 & 2 & -2 \\ 2 & 6 & 2 & 4 \\ -2 & 2 & -3 & 4 \\ 4 & 6 & 4 & 2 \end{pmatrix} . \qquad (10)$$

This is just the matrix of exponent sums for the group G as presented in the example in Section 1, sometimes called a *relation matrix* for G_{ab}. By performing row and column operations in accordance with the algorithm described below, M will later be reduced to a canonical form from which (i) free generators for B (see (5)) can be read off, and (ii) G_{ab} can be identified as a direct product of cyclic groups.

4.   The invariant factor theorem for matrices

Just as in ordinary linear algebra, pre- and post-multiplication by invertible matrices corresponds to performing elementary row and column operations:

P: permuting rows,

M: multiplying a row through by a unit (± 1),

A: adding to a row and scalar multiple of another row.

and similarly for columns. For obvious reasons and in an obvious way, these operations correspond to Nielsen transformations on sets of words in a free group ((T3′), (T1), (T2′), respectively: see Section 3.1). We now describe an algorithm for reducing any $r \times s$ matrix M over $\mathbb{Z}$ to the canonical form:

$$D = \text{diag}\,(d_1, d_2, ..., d_k), \quad k = \min(r,s) , \tag{11}$$

$$d_i \times \mathbb{N} \cup \{0\}, \ 1 \leq i \leq k, \ d_i \,|\, d_{i+1}, \ 1 \leq i \leq k-1 . \tag{12}$$

D is again an $r \times s$ matrix, and has all entries off the main diagonal equal to zero.

Remark 2. Note that the divisibility condition (12) implies that any 1's among the d's must occur at the beginning, and any 0's at the end.

We now give the steps of the algorithm in words, and then illustrate it by an example. Note the strong analogy with Gaussian elimination.

1.   Pick an entry of d of M of minimal positive modulus (if no such exists, then M is the zero matrix, which is already in canonical form) and remove it to the (1,1)-place by P-operations.

2.   Use an M-operation to ensure that $d > 0$.

3.   Divide the (2,1)-entry by d (according to Euclid), and use an A-operation to replace it by b, $0 \leq b < d$. (If $s = 1$, go at once to step 7).

4.   If $b = 0$, go to step 6, and if not transpose rows 1 and 2 and revert to step 3, with b in place of d.

5.   Repeat steps 3 and 4 until the (2,1)-entry is zero (whereupon the new d in the (1,1)-place will be the highest common factor of the (1,1)- and (2,1)-entries at the start of step 3).

6. Perform steps 3-5 on the remaining rows until the only non-zero entry in the first column is the d in the (1,1)-place.

7. Perform steps 3-6 on columns until every entry in the first row is zero except for the d in the (1,1)-place.

8. If d divides every entry in the matrix, go to step 11. If not, use a P-operation and an A-operation to get b into the (2,1) place with $d \nmid b$.

9. Repeat steps 3-5 until the only non-zero entry in the first column is the d in the (1,1)-place, then return to step 7 with the new d.

10. Repeat steps 8 and 9 until the (1,1)-entry $d = d_1$ divides every other entry in the matrix.

11. Apply steps 1-10 to the matrix obtained by removing the first row and column to obtain a d_2 in the (1,1)-place that divides every other entry (and is divisible by d_1).

12. Repeat step 11 until either rows or columns or non-zero entries run out.

Proposition 6. Given any matrix M over $\mathbb{Z}$, there exist invertible matrices T and Q^{-1} such that TMQ^{-1} has the canonical form D of (11), (12).

Remark 3. The d_i's so obtained are called the invariant factors of M. Their uniqueness is easily proved as follows. Let $h_i(M)$ denote the hcf. of the i-rowed minors of M ($1 \leq i \leq k$). Since the i-rowed minors (up to a sign), and hence their hcfs., are preserved by elementary row and column operations, we have

$$h_i(M) = h_i(D) = d_1 \ldots d_i, \quad 1 \leq i \leq k \ ,$$

so that the d_i's, being quotients of the $h_i(M)$, are determined by M.

Example. We apply the algorithm to the matrix M of (10). Note that the process described above for changing the generators of B remains valid even when the generators are not free, since the premultiplication of M by T corresponds to a regular Nielsen transformation, and thus yields a new set of generators. As the element of smallest positive modulus, pick the "2" in the (2,1)-place. Step 1 thus consists of transposing the first two rows to obtain:

$$\begin{pmatrix} 2 & 0 & 2 & -2 \\ 6 & 6 & 6 & 0 \\ 2 & 6 & 2 & 4 \\ -2 & 2 & -3 & 4 \\ 4 & 6 & 4 & 2 \end{pmatrix}.$$

Perform steps 2-6 to clear the first column (four row operations of type A):

$$\begin{pmatrix} 2 & 0 & 2 & -2 \\ 0 & 6 & 0 & 6 \\ 0 & 6 & 0 & 6 \\ 0 & 2 & -1 & 2 \\ 0 & 6 & 0 & 6 \end{pmatrix}.$$

Step 7 clears the first row:

$$\begin{pmatrix} 2 & 0 & 0 & 0 \\ 0 & 6 & 0 & 6 \\ 0 & 6 & 0 & 6 \\ 0 & 2 & -1 & 2 \\ 0 & 6 & 0 & 6 \end{pmatrix}.$$

For step 8, transpose rows 2 and 4; and then add column 3 to column 1:

$$\begin{pmatrix} 2 & 0 & 0 & 0 \\ -1 & 2 & -1 & 2 \\ 0 & 6 & 0 & 6 \\ 0 & 6 & 0 & 6 \\ 0 & 6 & 0 & 6 \end{pmatrix}.$$

Step 9 adds row 1 to row 2, then transposes the first two rows, then subtracts twice row 1 from row 2, and finally clears the first row (step 7):

$$\begin{pmatrix} 1 & 0 & 0 & 0 \\ 0 & -4 & 2 & -4 \\ 0 & 6 & 0 & 6 \\ 0 & 6 & 0 & 6 \\ 0 & 6 & 0 & 6 \end{pmatrix}.$$

Step 11 applied to 4×3 submatrix in lower right-hand corner yields

$$\begin{pmatrix} 2 & 0 & 0 \\ 0 & 6 & 6 \\ 0 & 6 & 6 \\ 0 & 6 & 6 \end{pmatrix}.$$

Finally, step 12 yields the 3×2 matrix

$$\begin{pmatrix} 6 & 0 \\ 0 & 0 \\ 0 & 0 \end{pmatrix}.$$

M is thus reduced to the canonical form

$$\begin{pmatrix} 1 & 0 & 0 & 0 \\ 0 & 2 & 0 & 0 \\ 0 & 0 & 6 & 0 \\ 0 & 0 & 0 & 0 \\ 0 & 0 & 0 & 0 \end{pmatrix}, \tag{13}$$

with d_1, d_2, d_3, d_4 respectively equal to 1,2,6,0. Denoting the new generators for A by x_1, x_2, x_3, x_4, the new generators for B (see (5)) are $\{x_1, 2x_2, 6x_3\}$, so that the rank of B is 3. The new coefficient matrix is obtained from (13) by deleting the last two rows (corresponding to two Nielsen transformations of type (TO)).

5.   The basis theorem

Translating the result of the previous section back into group theory, our arbitrary f.g. abelian group takes the form

$$\frac{A}{B} = \frac{A(\{x_1,...,x_r\})}{<d_1x_1,...,d_kx_k>} , \quad k \leq r ,$$

$$\cong \frac{<x_1|> \times ... \times <x_k|> \times ... \times <x_r|>}{<d_1x_1) \times ... \times <d_kx_k>} , \quad \text{by Proposition 2} ,$$

$$\cong \frac{<x_1|>}{<d_1x_1>} \times ... \times <x_k|> \times ... \times \frac{<x_k|>}{<d_kx_k>} \times ... \times <x_r|> , \quad \text{by Exercise 4} ,$$

$$\cong <x_1|x_1^{d_1}> \times ... \times <x_k|x_k^{d_k}> \times ... \times <x_r|>$$

$$\cong Z_{d_1} \times ... \times Z_{d_k} \times Z \times ... \times Z , \tag{14}$$

where Z_0 is to be interpreted as Z, and there are $r-k$ unsubscripted Z's.

In accordance with Remark 2,

(i)   any 1's among the d's occur at the beginning: discard these (trivial) factors

(ii) any 0's among the d's occur at the end: absorb these (infinite cyclic) factors into the Z's.

Having done this, and renumbered, the d_i's in (14) all belong to $\mathbb{N} \setminus \{1\}$, and the divisibility condition (12) still holds. These d's (called the *invariant factors* of A/B) and the number of Z's (called the *rank* of A/B) are unique in view of Remark 3.

Theorem 1 (The basis theorem). Given a finitely-generated abelian group G, there are integers $k,n \geq 0$ and integers $d_i \geq 2$, $1 \leq i \leq k$, each dividing its successor, such that

$$G \cong Z_{d_1} \times ... \times Z_{d_k} \times Z^{\times n} .$$

Moreover, k,n and the $d_i (1 \leq i \leq k)$ are all determined by G.

In terms of presentations, this may be restated as follows.

Corollary 1. Every *f.g.* abelian group has a unique presentation of the form

$$<x_1,...,x_r \mid x_1^{d_1},...,x_k^{d_k},C> ,$$

where $k \leq r$ and the d_1 satisfy the conditions of the theorem.

A second corollary paves the way for the next chapter.

Corollary 2. If $G = <X \mid R>$ is a finite presentation of a finite group G, then $|X| \leq |R|$.

Proof. To prove the contraposed assertion, assume that $|X| > |R|$. Then the exponent-sum matrix M (see Section 3) associated with this presentation has fewer rows than columns. The number k of invariant factors in the expression (14) for G_{ab} is thus strictly less than $|X| = r$. Hence, there is at least one infinite cyclic factor in (14), which implies that G_{ab}, and thus also G, is infinite.

Example. To identify G_{ab} in our original example, apply (14) to (13) to obtain

$$G_{ab} \cong Z_2 \times Z_6 \times Z .$$

Exercises

1. Let G be a group and $H \triangleleft G$. Prove that

   $$G/H \text{ is abelian} \Leftrightarrow H \supseteq G' .$$

2. Show that if $w \in F(X)$ is a word in which every $x \in X$ occurs with exponent-sum zero, then $w \in F(X)'$.

3. Prove from first principles that every non-trivial subgroup of the infinite cyclic group Z is infinite cyclic.

4. Let G, H be groups with $A \triangleleft G$, $B \triangleleft H$. Then prove that $A \times B \triangleleft G \times H$ and that

   $$\frac{G \times H}{A \times B} \cong \frac{G}{A} \times \frac{H}{B} .$$

5. Given an abelian group A, put

   $$\tau(T) = \{g \in G \mid g \text{ has finite order}\} .$$

   Prove that $\tau(A)$ is a subgroup of A (it is called the *torsion subgroup*). In the case when A is $f.g.$, show further that

   (i) $\tau(A)$ is finite, (ii) $A/\tau(A)$ is a $f.g.$ free abelian group.

6. Let $G = <X \mid R>$ be a finite presentation of a finite group. Assume that $|X| = |R|$, so that the exponent-sum matrix M (see (9)) is a square matrix. Prove that

   $$|G| = \pm \det M .$$

7. Find the rank and invariant factors of G_{ab} for each of the following groups:

   (i)   the group Q_{2n} of Exercise 5.1,

   (ii) the Heisenberg group H of (5.2),

   (iii) the symmetric group S_n of (5.5),

   (iv) the metacyclic group of (5.14),

(v)   the group Sym L of Section 5.5,

(vi)   the polynomial group of (5.18),

(vii) the class 2 group of Exercise 5.5,

(viii) the group of order 16 in Exercise 5.14,

(ix)   the von Dyck group $D\,(l,m,n)$ of Example 4.5,

(x)   the knot group T of Example 4.6.

8.   Compute the order of G_{ab} when G is the group

$$<x,y,z\mid x^4 = y^{b-2}x^{-1}y^{b+2},\ y^z = z^{c-2}y^{-1}z^{c+2},\ z^x = x^{a-2}z^{-1}x^{a+2}>\ .$$

9.   Consider the abelianized Fibonacci groups

$$F\,(2,k)_{ab} = <x_1,...,x_k\mid x_1x_2 = x_3,...,x_kx_1 = x_2,\ C>\ .$$

Compute the invariant factors of this group for some small values of n. Using the generating set $\{x_1,x_2\}$, find a 2×2 coefficient-sum matrix for $F\,(2,k)_{ab}$ in terms of the Fibonacci numbers

$$1,1,2,3,5,8,13,21,...\ .$$

Find a formula for $|F\,(2,k)_{ab}|$ in terms of the Lucas numbers

$$1,3,4,7,11,18,...\ .$$

10.   Let him that hath understanding find the order of the group

$$B = <x,y,z,t \mid x^3y^7 = y^4x^7 = z^3t^5 = t^4z^6 = [x,z] = e>\ .$$

[Hint: it is the number of a man.]

CHAPTER 7

FINITE GROUPS WITH FEW RELATIONS

It has just been shown (Corollary 6.2) that if $G = <X \mid R>$ is a finite group, then $|X| \leq |R|$, so that finite groups have non-positive deficiency, in the following sense.

Definition 1. The *deficiency* of a finite presentation $<X \mid R>$ is defined to be $|X| - |R|$. The *deficiency* of a finitely-presented group G is given by $\mathrm{def}\, G = \max \{ |X| - |R| \mid$ finite presentations $<X \mid R>$ of $G\}$.

Definition 2. For a finite group $G = <X \mid R>$, the (Schur) *multiplicator* of G is defined by

$$M(G) = \frac{F' \cap \bar{R}}{[F, \bar{R}]} \; ,$$

where $\bar{R}$ is the normal closure of R in $F = F(X)$.

In 1907, J. Schur proved (among other things) that $M(G)$ is

(i)   an *invariant* of G, that is, independent of the finite presentation $<X \mid R>$,

(ii)   a *finite* abelian group,

(iii)   generated by $-\mathrm{def}\, G$ elements.

Though light will be shed on these and other results later, their proofs are beyond our present scope. Our purpose here is to give examples of finite groups of deficiency zero, hereafter referred to as *interesting* groups. From what has been said, interesting groups have trivial multiplicator, but not conversely. The class of finite groups with trivial multiplicator but non-zero deficiency was shown by R.G. Swan to include some soluble groups; whether it contains any nilpotent groups is an unsolved problem.

The interesting groups exhibited below are roughly classified into various types. Note that no example is known of an interesting group that needs more than three generators.

However, it follows from the Golod-Shafarevich theorem (see Chapter 15, below) that all interesting nilpotent groups are at most 3-generated.

1.   Metacyclic groups

The simplest examples of interesting groups are, of course, the finite cyclic groups (see Remark 4.4). Other examples are the generalised quaternion groups Q_{2^n} (see Exercises 5.1 and 2, below) and certain of the metacyclic groups (5.14) (see Exercise 3, below).

Definition 3. A group G is called metacyclic if it has a normal subgroup H such that both H and G/H are cyclic.

In the case where G is finite, we can thus assume that

$$H = <x> \cong Z_m, \ G/H = <Hy> \cong Z_n , \tag{1}$$

with $x,y \in G$, $m,n \in N$. Since H is a normal subgroup of index n, both $y^{-1}xy$ and y^n must belong to H, say

$$y^{-1}xy = x^r, \ y^n = x^s , \tag{2}$$

where r,s are integers with $1 \le r, \ s \le m$. Now it follows from the first relation of (2) that for $a,b \in \mathbb{Z}, b \ge 0$,

$$y^{-b}x^a y^b = x^{ar^b} , \tag{3}$$

(see Exercise 1), so that the second relation of (2) implies that

$$x = x^{-s}xx^s = y^{-n}xy^n = x^{r^n} ,$$

whence, $r^n \equiv 1 \pmod m$, since $|x| = |H| = m$. Similarly, we have

$$x^s = y^n = y^{-1}y^n y = y^{-1}x^s y = x^{rs} ,$$

so that $rs \equiv s \pmod m$. We are now in a position to write down presentations for all finite metacyclic groups.

Proposition 1. Consider the group

$$G = <x,y \, | \, x^m = e, \ y^{-1}xy = x^r, \ y^n = x^s> , \tag{4}$$

where

$$m,n,r,s \in N, \ r, s \le m,$$

and

$$r^n \equiv 1, \ rs \equiv s \ (\text{mod } m). \tag{5}$$

Then $N = <x>$ is a normal subgroup of G such that

$$N \cong Z_m, \ G/N \cong Z_n.$$

Thus, G is a finite metacyclic group, and moreover, every finite metacyclic group has a presentation of this form.

Proof. To prove that $N \triangleleft G$, write any member w of G as a word in x,y. Since $x^m = e$ and $y^n = x^s$, we can write w as a positive word and consider it as an alternating product of syllables of the form x^i, y^j with $i, j \in N$. Now conjugation of any power of x by either of these again yields a power of x, by the consequence (3) of the relation $y^{-1}xy = x^r$. Hence, $N^w \subseteq N$ for all $w \in G$ and $N \triangleleft G$. It now follows from Proposition 4.2 that G/N is given by adjoining the relation $x = e$ to the presentation (4), and this plainly yields Z_n.

While we know already that N is a factor group of Z_m, it is a non-trivial matter to prove that these groups are actually isomorphic. To do this, consider the set C of ordered pairs (i,j) where i is an integer with $0 \le i \le n-1$ and j is a residue class modulo m. We define a binary operation on G by setting

$$(i,j)(k,l) = \begin{cases} (i + k, \ l + jr^k) & , \text{ if } i + k < n, \\ (i + k - n, \ l + jr^k + s) & , \text{ if } i + k \ge n, \end{cases} \tag{6}$$

where r and s are as in the statement of the theorem. We claim that C is a group, first observing that $(0,0)$ is an identity, $(0,j)^{-1} = (0,-j)$, and for $i > 0$, it follows from (6) that

$$(i,j)^{-1} = (n - i, \ -jr^{n-i} - s).$$

Checking the associative law is rather tedious, and we merely observe that because of (5), the product of (i,j), (k,l), (a,b) is equal to

$$(i + k + a, \ b + lr^a + jr^{k+a}),$$

$$(i + k + a - n, \ b + lr^a + jr^{k+a} + s),$$

$$(i + k + a - 2n, \ b + lr^a + jr^{k+a} + 2s),$$

according as $[(i + k + a)/n] = 0,1,2$ respectively, and is thus independent of the bracketing. Now in C,

$$(i,l) = (i, 0)(0,l) = (1,0)^i (0,1)^l ,$$

so that $(0,1)$ and $(0,1)$ generate C. Substitution of these for y and x respectively in the relations of (4) yields identities, so that C is a factor group of G by the substitution test. Hence C is a concrete realization of G, and $|N| = m$ as required. The last part of the theorem has already been proved.

While this theorem gives a description of any finite metacyclic group in terms of the four parameters m,n,r,s, it is *not* a classification theorem. In fact, the problem of when two groups of the type (4) are isomorphic is unsolved in general, though the special case when mn is a prime-power has recently been dealt with by F.R. Beyl.

We see from the presentation (4) that def G is either -1 or 0 and so $M(G)$ must be cyclic by the classical results of Schur mentioned in the introduction of this chapter, and by the same token, $M(G)$ is trivial if def $G = 0$. That the converse holds for finite metacyclic groups was shown by J.W. Walmsley in 1970, who gave a 2-generator 2-relation presentation when $M(G) = E$. A slicker version was given three years later by Beyl, and we describe this now.

There are two problems involved:

(i)   to compute $|M(G)|$ in terms of m,n,r,s, and

(ii) to define G by two relations when $M(G) = E$.

We shall give a paraphrase of (i) and prove (ii) in full (Proposition 2 below). To get at $M(G)$, we need to take a closer look at the congruences (5). From the first of these, it follows that

$$h = \frac{1}{m}(m,r-1)(m, 1 + r + \ldots + r^{n-1})$$

is an integer. From the second congruence, there is an integer k such that

$$s = \frac{km}{r-1} = \frac{km/(m,r-1)}{(r-1)/(m,r-1)} = lm/(m,r-1) ,$$

where l is an integer since $m/(m,r-1)$ and $(r-1)/(m,r-1)$ are coprime. By changing generators and using elementary number theory, l can be replaced by (l,h), so that we can take $s = lm/(m,r-1)$ in (4), with l a divisor of h. Beyl's main result now asserts that l is nothing other than $|M(G)|$. Granting this, the deficiency problem for finite metacyclic groups is solved as follows.

Proposition 2. Let

$$G = \langle x, y \mid x^m = e, \; y^{-1}xy = x^r, \; y^n = x^s \rangle,$$

where $m, n, r, s \in N$ and

$$r^n \equiv 1 \pmod{m} \quad \text{and} \quad s = m/(m, r-1).$$

Then G is presented, in terms of the same generators x and y, by

$$\langle x, y \mid y^n = x^s, \; [y, x^{-t}] = x^{(m, r-1)} \rangle, \tag{7}$$

where t is a certain integer.

Proof.

(i) We first construct an integer t such that the second relation of (7) holds in G. To this end, let

$$(m, r-1) = u(r-1) + vm,$$

so that $(u, s) = 1$. Now let w be the largest factor of m coprime to u, and put $t = u + ws$. Now modulo m,

$$(m, r-1) \equiv u(r-1) \equiv t(r-1),$$

since $s(r-1)$ is divisible by m. Hence, in G,

$$[y, x^{-t}] = (y^{-1}x^t y) x^{-t} = x^{(r-1)t} = x^{(m, r-1)},$$

and the relations of (7) hold in G.

(ii) We now assume the relations of (7) and show that they define G. Armed with the knowledge that x^s commutes with y and $[y, x^{-t}]$ commutes with x, we see that

$$x^{ts} = y^{-1}x^{ts}y = (y^{-1}x^t y)^s = ([y, x^{-t}]x^t)^s = [y, x^{-t}]^s x^{ts},$$

and so

$$x^m = x^{(m, r-1)s} = [y, x^{-t}]^s = e.$$

To complete the proof, we first need to show that t and m are coprime, so assume (for a contradiction) that p is a prime factor of both. Since p divides m, it divides exactly one of w and u. But p divides $t = u + ws$, so it cannot divide w. Hence, p divides u and s contrary to the fact that $(u, s) = 1$. Thus, there is an integer k such that $kt \equiv 1 \pmod{m}$, and we have

$$[y,x^{-1}] = [y,x^{-kt}] = ([y,x^{-t}]x^t)^k x^{-kt} = [y,x^{-t}]^k$$
$$= x^{k(m,r-1)} = x^{kt(r-1)} = x^{r-1}$$

Hence $x^y = x^r$, and the three relations of G follow from the two relations of (7).

2.   Interesting groups with three generators

Thirty years ago, the list of known interesting groups comprised

(i)   cyclic groups,

(ii)   certain metacyclic groups (see §1) studied by Schur,

(iii) certain derivatives of the von Dyck groups (see Example 4.5) studied by G.A. Miller.

Now all these groups are 2-generated, and the question naturally arose as to whether there were any interesting groups needing 3 generators or more. The first examples were provided by J. Mennicke in 1959, who showed that the groups

$$M(a,b,c) = <x,y,z \mid y^{-1}xy = x^a, z^{-1}yz = y^b, x^{-1}zx = z^c>$$

and finite in the case $a = b = c \geq 3$. Eleven years later, Wamsley showed that the two classes of groups

$$W_\pm(a,b,c) = <x,y,z \mid x^z = x^a, y^{z^{\pm 1}} = y^b, z^c = [x,y]>$$

were interesting provided $(a-1)(b-1)c \neq 0$, this work being based on the groups Mac$(a,b) = W_-(a,b,1)$ discovered by I.D. Macdonald in 1962. While our current stock of interesting groups with 2-generators is very large, the list of those needing 3 generators is fairly short. It consists of 3-generator groups in the above three classes of Mennicke and Wamsley, together with the groups

$$J(a,b,c) = <x,y,z \mid x^y = y^{b-2}x^{-1}y^{b+2}, \, y^z = z^{c-2}y^{-1}z^{c+2}, \, z^x = x^{a-2}z^{-1}x^{a+2}> ,$$

where a,b,c are non-zero *even* integers (see Exercise 6.8), and some generalizations of the $M(a,b,c)$ discovered by M.J. Post. Since the $J(a,b,c)$ are the easiest to analyse, we investigate this class now, observing that similar methods apply to the other classes.

Let $G = J(a,b,c)$. The first step is to conjugate the first relation by y:

$$x^{y^2} = y^{b-2}(x^y)^{-1}y^{b+2} = y^{-4}xy^4 = x^{y^4} ,$$

so that x commutes with y^2 (write $x \sim y^2$). From the other two relations it follows that

$$y \sim z^2, \, z \sim x^2 , \tag{8}$$

and thus that the subgroup $H = \langle x^2, y^2, z^2 \rangle$ of G is abelian. Furthermore, the original relations imply that

$$x^y = x^{-1}y^{2b}, \ y^z = y^{-1}z^{2c}, \ z^x = z^{-1}x^{2a}, \tag{9}$$

and as a consequence,

$$(x^2)^x = x^2, \ (x^2)^y = x^{-2}y^{4b}, \ (x^2)^z = x^2$$

all belong to H. Since the conjugates of x^2 by x^{-1}, y^{-1}, z^{-1} are equal to its conjuages by x, y, z respectively (H is abelian), we see that $(x^2)^w \in H$ for any $w \in G$. Similarly, $(y^2)^w, (z^2)^w \in H$ and so H is *normal* in G. By Proposition 4.2 a presentation for G/H is given by adjoining the relations $x^2 = y^2 = z^2 = e$ to those defining G, and after a couple of Tietze transformations, we have

$$G/H = \langle x, y, z \mid x^2 = y^2 = z^2 = (xy)^2 = (yz)^2 = (zx)^2 = e \rangle .$$

Thus, $G/H \cong Z_2 \times Z_2 \times Z_2$ and $|G:H| = 8$.

Now apply the relations (8) and (9) together with the fact that $x^2 \sim y^2$ to compute that

$$[[x,y], z^m] = [x^{-2}y^{2b}, z^{-1}x^{2a}] = [y^{2b}, z^{-1}] = [y^{2b}, z] = y^{-2b}(y^z)^{2b}$$

$$= y^{-2b}(y^{-1}z^{2c})^{2b} = y^{-4b}z^{4bc} .$$

Substituting this and its companions into the Witt identity (see Exercise 4.13),

$$e = [[x,y], z^x][[z,x], y^z][[y,z], x^y]$$

$$= y^{-4b}z^{4bc} \cdot x^{-4a}y^{4ab} \cdot z^{-4c}x^{4ca}$$

$$= x^{4a(c-1)}y^{4b(a-1)}z^{4c(b-1)},$$

since x^2, y^2, z^2 all commute. Since y^2, z^2 commute with y so does $x^{4a(c-1)}$, and thus

$$x^{4a(c-1)} = (x^{4a(c-1)})^y = (x^{-1}y^{2b})^{4a(c-1)} = x^{-4a(c-1)}y^{8ab(c-1)} .$$

Hence

$$x^{8a(c-1)} = y^{8ab(c-1)},$$

and similarly,

$$y^{8b(a-1)} = z^{8bc(a-1)}, \ z^{8c(b-1)} = x^{8ca(b-1)} .$$

Using each of these three relations, we finally obtain:

$$x^{8a(c-1)(a-1)(b-1)} = y^{8ab(c-1)(a-1)(b-1)}$$

$$= z^{8abc(c-1)(a-1)(b-1)}$$

$$= x^{8a^2bc(c-1)(a-1)(b-1)} ,$$

showing that x has order dividing

$$|8a(c-1)(a-1)(b-a)(abc-1)| , \tag{10}$$

which is non-zero. Thus x^2 has finite order and similarly, so do y^2 and z^2. Since x^2, y^2, z^2 all commute, it follows that $H = <x^2, y^2, z^2>$ is finite. Since $G/H = Z_2 \times Z_2 \times Z_2$ needs 3 generators, so does G, and we have the following result.

Proposition 3. Let G be the group

$$<x,y,z \mid x^y = y^{b-2}x^{-1}y^{b+2}, \ y^z = z^{c-2}y^{-1}z^{c+2}, \ z^x = x^{a-2}z^{-1}x^{a+2}> ,$$

where a,b,c are non-zero even integers, and let $H = <x^2, y^2, z^2> \leq G$. Then H is an abelian normal subgroup of G with G/H elementary abelian of order 8. The order of G is a divisor of

$$512 |abc|.|(a-1)(b-1)(c-1)(abc-1)|^3 . \tag{11}$$

The bound for $|G|$ is obtained by multiplying by 8 the product of the bounds for $|x^2|$, $|y^2|$, $|z^2|$ obtained from (10) and its analogues. Even in the simplest case, namely when $a = b = c = 2$, this yields a 7-digit number, while the order of G in this case is 7.2^{11} (the bound obtained by Wamsley). We must defer until a later chapter the method for computing $|G|$ exactly (see Exercise 9.15), merely observing that the correct value is

$$|G| = 2^8 |abc(abc-1)| . \tag{12}$$

We have proved that G is a metabelian group, that is, a group with abelian derived group. Recall that a group is called nilpotent if it has a finite chain of normal subgroups

$$G = G_1 > G_2 > ... > G_{n+1} = E$$

with $G_i/G_{i+1} \subseteq Z(G/G_{i+1})$ for all i. The class of a nilpotent group is the least value of n for which such a chain exists, so that for example, nilpotent groups of class 1 are just abelian groups. Simple commutator calculations show that $J(a,b,c)$ is nilpotent if and only

if $|abc|$ is a 2-power, whereupon its class is equal to $3 + \log_2 \max \{ |a|, |b|, |c| \}$.

3. Cyclically-presented groups

Cyclically presented groups comprise a potentially rich source of interesting groups, and indeed we have already looked at several examples of this type of group. We now give a formal definition, though the name is self-explanatory.

Definition 4. Let $F = <x_1,...,x_n|>$ and let θ be the automorphism of F induced by permuting the subscripts of the free generators in accordance with the cycle $(1\ 2\ ...\ n) \in S_n$. For any reduced word $w \in F$, the *cyclically-presented group* $G_n(w)$ is given by

$$G_n(w) = <x_1, x_2,...,x_n | w, w\theta, ..., w\theta^{n-1}> \ .$$

Since cyclically presented groups have non-negative deficiency, $G_n(w)$ is interesting if and only if it is finite. Examples of such groups appearing in the previous section are

$$M(a,a,a) = G_3(x_2^{-1}x_1x_2x_1^{-a}),$$

$$J(a,a,a) = G_3(x_1^{-1}x_2^{a-1}x_1^{-1}x_2^{a+1}),$$

$$\text{Mac}\ (a,a) = G_2(x_1^{[x_1,x_2]}x_1^{-a}).$$

Further examples are provided by the Fibonacci groups $F(2,n) = G_n(x_1x_2x_3^{-1})$ (cf. Exercise 6.9), and these are a special case of the following class.

Definition 5. For $r, n \in N$ with $r \geq 2$, the *Fibonacci group* $F(r,n)$ is defined as the cyclically-presented group

$$F(r,n) = G_n(x_1,...,x_rx_{r+1}^{-1}),$$

where subscripts are understood to be reduced modulo n to lie in the set $\{1,2,...,n\}$.

Table 3 gives an idea of what the $F(r,n)$ look like for small values of r and n. In the (r,n) place is written the order of $F(r,n)$, or the isomorphism type where appropriate. The gaps in the table correspond to gaps in our knowledge. We have already computed some of these entries and others figure among the next set of exercises. The hardest to identify is probably Campbell's group $F(3,6)$ - one of a bewildering array of cyclically presented groups of order 1512 studied by C.M. Campbell and E.F. Robertson - and this is the only finite entry in the table which is not metacyclic.

r	n=1	2	3	4	5	6
2	E	E	Q	Z_5	Z_{11}	∞
3	Z_2	Q	Z_2	∞	Z_{22}	1512
4	Z_3	Z_3	63	Z_3	∞	
5	Z_4	24	∞	624	Z_4	∞
6	Z_5	Z_5	Z_5	125	7775	Z_5
7	Z_6	48	342	∞		117648

Table 3

It is known that for fixed r, the $F(r,n)$ are eventually infinite; it is a nice exercise in small cancellation theory to prove that $F(r,n)$ is infinite for $n > 5r$. The case $r = 2$ is thus almost completely decided. A machine implementation of Todd-Coxeter coset enumeration (see Chapter 8) shows that $F(2,7) \cong Z_{29}$, while A.M. Brunner has proved that $F(2,8)$ and $F(2,10)$ are both infinite (see Exercise 26 and Exercise 10.). This leaves one group, namely $F(2,9)$, and the best we can say about this is as follows. By using a formidable array of algorithms and a considerable amount of ingenuity, G. Havas, J.S. Richardson, L.S. Sterling, a Univac 100/42 and a DEC KA 10 have proved that $F(2,9)$ has order at least 152.5^{741}.

Elementary methods (involving circulant matrices, resolvents of polynomials, and properties of complex roots of unity) yield a substantial amount of information about the groups $G_n(w)_{ab}$. We omit the details, and merely give examples of the type of result that can be obtained.

(i)   There is a formula for $|G_n(w)_{ab}|$ in terms of a certain polynomial associated with w.

(ii)   Precise conditions can be given for $G_n(w)$ to be (a) trivial, and (b) infinite.

(iii) $F(r,n)_{ab}$ is always a finite group.

(iv)   For a fixed r, $|F(r,n)_{ab}| < |F(r,n+1)_{ab}|$ when n is sufficiently large.

(v)   Given a group G, there are at most finitely many pairs (r,n) such that $G \cong F(r,n)$.

(vi)   On the other hand, $\forall\, k \in \mathbb{N}$, $\exists$ a group G such that G appears at least k times among the $F(r,n)$.

Proposition 4. Every finite group is a homomorphic image of some $F(r,n)$.

Proof. Let G be a finite group, with generators $x_1, x_2, ..., x_r$, say. For $k \in \mathbb{N}$, make the inductive definition

$$x_{r+k} = x_k x_{k+1} \cdots x_{r+k-1} ,$$

and consider the sequence of r-tuples

$$v_k = (x_k, ..., x_{r+k-1}) .$$

Since these all belong to the finite set $G^{\times r}$, two of them are equal, say $v_m = v_{m+n}$, $m,n \in \mathbb{N}$. As any set of r consecutive x's determine their immediate predecessor, it follows from this that $v_1 = v_{n+1}$. The defining relations of $F(r,n)$ thus hold in G for $x_1, ..., x_n$, whence $G = <x_1, ..., x_n>$ is a homomorphic image of $F(r,n)$, by Proposition 4.3.

Exercises

1. Let x and y be members of a group G such that

    $$y^{-1}xy = x^r, \ r \in \mathbb{Z} .$$

    Prove that for $a,b \in Z$ with $b \geq 0$,

    $$y^{-b}x^a y^b = x^{ar^b} .$$

2. Prove that one of the relations in the presentation for Q_{2n} given in Exercise 5.1 is superfluous.

3. Use Exercise 1 to show that the metacyclic group presented in (5.14) is interesting in the case when $l - 1$ and $(l^m - 1)/n$ are coprime.

4. Consider the 'archetypal metacyclic group'

    $$G = <x,y \mid y^{-1}xy = x^r> .$$

    Prove that every element in G can be written in the form

    $$y^i x^j y^{-k}, \ i,j,k \in \mathbb{Z}, \ i,k \geq 0 .$$

    Identify the elements of G', and deduce that G' is abelian. Do you believe that G really is metacyclic?

5. Suppose that

$$G = <x,y \mid x^m = e, \ x^y = x^r, \ y^n = x^r > .$$

Use Beyl's criterion to prove that $M(G)$ is cyclic of order

$$(m, r - 1)(m, \ 1 + r + \dots + r^{n-1})/m .$$

6. Use the result of the previous exercise to compute the multiplicator of the dihedral group D_m of degree m. Write down a 2-generator 2-relation presentation of D_m when m is odd.

7. Compute $M(Z_m \times Z_n)$, $m,n \in N$.

8. Let p be an odd prime, $a,b \in \mathbb{N}$ and $k \in \mathbb{Z}$ such that $(k,p) = 1$. Prove that

$$(1 + kp^a)^{p^b} = 1 + kp^{a+b} + lp^{a+b+1} ,$$

    with $l \in \mathbb{Z}$.

9. Use Exercise 8 to find a deficiency-zero presentation of the group

$$G = <x,y \mid x^{p^{a+b}} = e, \ y^{-1}xy = x^{1+p^a}, \ y^{p^b} = e >$$

    when p is an odd prime and $a,b \in \mathbb{N}$.

10. Prove that the groups G of Exercise 9 exhaust the class of split metacyclic groups with odd prime-power order and trivial multiplicator.

11. Prove that the group

$$x,y \ x^m = y^n = [x,y]$$

    is cyclic when $(m,n) = 1$.

12. Use (12) to compute the order of $J(a,b,c)'$ (see Exercise 6.8).

13. By studying relations derived in the proof of Proposition 3 try to reduce as far as possible the bound (11) for the order of $J(a,b,c)$.

14. When do abelianized Mennicke and Walmsley groups need 3 generators? Write down necessary and sufficient conditions on a,b,c in each case.

15. Identify $M(a,a,a)$ when $a = 0,1,2$.

16. Use the Witt identity to prove that x,y,z each have finite order in $M(a,b,c)$ when $|a - 1|, |b - 1|, |c - 1|$ are all at least 2.

Exercises

17. Prove that every element of $M(a,b,c)$ can be written in the form $x^i y^i z^k$, $i,j,k \in \mathbb{Z}$. Deduce that $M(a,b,c)$ is finite when $|a-1|$, $|b-1|$, $|c-1|$ are all at least 2, and write down a bound for the order in this case.

18. Find a finite group that needs 4 generators and can be defined by 6 relations.

19. Prove that $x_n \ldots x_1 = e$ in $F(2,n)$. Show further that $x_1 \ldots x_n = e$ or $(x_1 \ldots x_n)^2 = e$ according as n is even or odd.

20. Prove that $[x_1, x_2]^2 = e$ in $F(2,n)$.

21. Prove that $F(n, n+1) \cong F(2n-1, n)$ for all $n \geq 2$.

22. Show that $F(r,n) \cong \mathbb{Z}_{r-1}$ when n is a divisor of r.

23. Show that the cyclic group of order 2^{s-1} appears at least s times among the $F(r,n)$.

24. Show that for all $s \geq 1$, $F(2s+1, 2)$ is a metacyclic group of order $4s(s+1)$.

25. Prove that when $r \equiv 1 \pmod n$, $F(r,n)$ is a metacyclic group of order at most $n(r^n - 1)$.

26. Use the matrices

$$A_1 = \begin{pmatrix} 1 & 1 & 0 \\ -1 & 0 & 0 \\ 0 & 0 & 1 \end{pmatrix}, \quad A_2 = \begin{pmatrix} 0 & 1 & 0 \\ 0 & 0 & 1 \\ 1 & 0 & 0 \end{pmatrix}.$$

to show that the group $F(2,10)$ is infinite.

27. Show that $F(2,14)$ is infinite (as in Exercise 4.11) using the permutations

$$
\alpha_1 : \begin{array}{ccc} \mathbb{Z} & \rightarrow & \mathbb{Z} \\ 2n & \mapsto & 2n+2 \\ 2n+1 & \mapsto & 2n-1 \end{array}, \quad
\alpha_2 : \begin{array}{ccc} \mathbb{Z} & \rightarrow & \mathbb{Z} \\ 3n & \mapsto & 3n-3 \\ 3n+1 & \mapsto & 3n+2 \\ 3n+2 & \mapsto & 3n+4 \end{array}.
$$

CHAPTER 8

COSET ENUMERATION

The method known as "systematic enumeration of cosets" was first propounded in 1936 by J.A. Todd and H.S.M. Coxeter. It works in any specific situation when H is the subgroup of a group $G = <X \mid R>$ generated by a set Y of words in $X^{\pm 1}$, provided only that $|X|$, $|R|$, $|Y|$ and $|G : H|$ are all finite. It is a purely mechanical process that starts with input X, R, Y, and ends with information about G relative to H. The history of its machine implementation goes back to the very early days of computers, and has developed with them up to the present time, when it comprises, in original or modified form, an important part of the many group theory packages, such as CAYLEY, in current use on machines throughout the world.

1. The basic method

The method described below can be applied to any *specific* finite presentation $G = <X \mid R>$, and always works provided $|G|$ is finite. It not only yields such information as

the order of G,

a faithful permutation representation of G,

a Cayley diagram for G, and

a Schreier transversal for $\overline{R}$ in $F(X)$,

but is also great fun. This is how it goes.

For each relation $r = x_1 \ldots x_n \in R$, with $x_1 \ldots x_n$ a reduced word in $X \cup X^{-1}$, we draw a rectangular table having $n + 1$ columns and a certain (for the moment unspecified) number of rows:

	x_1	x_2	...	x_n
1	2		...	1
2			...	2

Fig. 11

We begin by entering the symbol 1 in the first and last places of the first row of each table, the remaining places in the first rows being as yet empty. We then pick an empty space next to some 1 (either to the right or left of it) and fill it with the symbol 2. For the sake of definiteness, suppose the situation to be as in the above diagram, with 2 immediately to the right of 1 and $x_1 \in X \cup X^{-1}$ lying between them. We record the information '$1x_1 = 2$' (and / or '$2x_1^{-1} = 1$') in a monitor table: such an equation is eminently reasonable if we think of the numbers 1,2 as corresponding to the elements $e, x_1 \in G$, respectively. Now we put a 2 in the first and last places of the second row of each table and, wherever in any table 1 lies to the left of an empty space with x_1 between the two spaces, or to the right of an empty space with x_1^{-1} between, we fill that empty space with a 2. Similarly, if 2 lies to the right (left) of an empty space with $x_1 (x_1^{-1})$ between, we fill that space with a 1. This purely mechanical process is known as *scanning*, and it is here that most of the work is involved. Having made sure that no more spaces can be filled in this way, we enter the symbol 3 in any empty space that is adjacent to a filled space. Having recorded the corresponding information (of the form, '$ix = 3$') in our monitor table, we begin a new row in the relator tables and scan as above. In similar fashion, we introduce the symbol 4, record information, begin a fourth row, and scan again. We continue in this way until there are no more empty spaces, whereupon the number of rows in each relator table is equal to $|G|$.

In order to reach the situation where all the tables are complete, we clearly need more information (of the form $ix = j$) than is contained in our definitions of new symbols, and this is supplied when any row of any table becomes complete. For suppose we are in the position where such a row has but one remaining empty space, and that space is filled as indicated in Figure 12:

$$(1)$$

Fig. 12

Now this transition involves two pieces of information, namely

$$ix_l = j, \quad kx_{l+1}^{-1} = j \ ,$$

and we must distinguish between two cases. If this is the first time the symbol j has appeared, one of these two equations is a definition, and the other may be regarded as a bonus. On the other hand, the arrival of j may be the result of scanning when either one or both of these equations is already known. The latter case yields no new information, while in the former we obtain a bonus as before. There is one further possibility, but we postpone this in order to give a couple of examples.

Example 1. Take the cyclic group $G = <x \mid x^4>$ of order 4. Here, there is only one relator table; it is headed $xxxx$ and has five columns:

	x		x		x		x	
1	–	2	–	3	–	4	≡	1
2		3		4	≡	1		2
3		4	≡	1		2		3
4	≡	1		2		3		4

Table 4

We have made the definitions

$$1x = 2, \ 2x = 3, \ 3x = 4 \ ,$$

and with the completion of the first row, have obtained as a bonus that $4x = 1$. For explanatory purposes, we have indicated this by dashes on the vertical lines of the table at the corresponding points - one dash for a definition, two for a bonus, and three when a row completes without yielding new information.

Since the table completes after 4 rows, we deduce that $|G| = 4$. Had we bothered to draw them, our monitor tables would have looked like this:

definition	bonus
$1x = 2$	
$2x = 3$	
$3x = 4$	$4x = 1$

	x
1	2
2	3
3	4
4	1

Table 5

The Schreier transversal comes from the "definition" column of the first monitor table. Its jth element is the word u_j in $X^{\pm}$ which defines the coset $j = 1u_j$ (recursively) in terms of the coset 1, and is $\{e,x,x^2,x^3\}$ in this case. The second monitor table yields the (regular) permutation representation $x \mapsto (1234) \in S_4$, and the corresponding Cayley diagram is as follows:

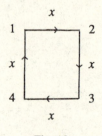

Fig. 13

Example 2. As an example where the answer is not quite so obvious in advance, consider the Fibonacci group

$$G = F(2,3) = <a,b,c \mid ab = c, \ bc = a, \ ca = b> .$$

Eliminating c by Tietze transformations, we have

$$G = <a,b \mid baba^{-1}, \ abab^{-1}> ,$$

and thus two relator tables. The reader can either check the working, or perferably do the enumeration himself using only the 'definition' column in the first monitor table below. If the same definitions are made in the same order, the result should look like this:

	b	a	b	a⁻¹
1	– 2	– 3	– 4	= 1
2	– 6	– 8	= 3	2
3	4	6	= 7	– 3
4	5	2	6	4
5	8	1	2	5
6	7	5	8	6
7	1	4	5	7
8	3	7	1	8

	a	b	a	b⁻¹
1	4	– 5	= 2	1
2	3	4	= 6	2
3	7	= 1	4	3
4	6	7	= 5	4
5	2	6	8	= 5
6	8	3	7	6
7	5	8	1	7
8	= 1	2	3	8

Table 6

It follows that $F(2,3)$ has order 8. We have omitted the triple dashes here; as with the double dashes, the points where they appear are not always unique. The monitor tables are as in Table 7:

definition	bonus			a	b
$1b = 2$			1	4	2
$2a = 3$			2	3	6
$3b = 4$	$1a = 4$		3	7	4
$4b = 5$	$5a = 2$		4	6	5
$2b = 6$	$4a = 6$		5	2	8
$3a = 7$	$6b = 7$		6	8	7
	$7b = 1$		7	5	1
	$7a = 5$		8	1	3
$6a = 8$	$8b = 3$				
	$5b = 8$				
	$8a = 1$				

Table 7

Returning to the general case, we continue the analysis of what occurs when a row completes, where the following alarming contingency remains to be considered. It may (and sometimes does) happen that in the process of scanning, the bonus information obtained from the completion of a row is inconsistent with what we already know. In terms of Fig. 12, this occurs when our monitor tables at this point contain the information

$$ix_l = j, \ kx_{l+1}^{-1} = m \qquad\qquad j \neq m ,$$
$$\text{or} \quad ix_l = j, \ jx_{l+1} = n \qquad \text{with} \quad k \neq n ,$$
$$\text{or} \quad jx_l^{-1} = h, \ kx_{l+1}^{-1} = j \qquad \text{with} \quad h \neq i ,$$
$$\text{or} \quad ix_l = m, \ kx_{l+1}^{-1} = j \qquad \text{with} \quad j \neq m .$$

In every case we obtain inconsistent information of the form $ix = j$, $ix = m$ (say), with $j \neq m$. This induces the phenomenon known as *coset collapse*, and we must proceed as follows. We conclude that $j = m$, and replace the larger of these numbers by the smaller throughout *all* our tables, noting that this converts the definition of the larger symbol into a bonus. This may yield extra bonuses treated in the usual way or further collapses, whereupon we pursue the same strategy, and continuing in this way until all inconsistent information has been removed. We then delete all rows in the relator tables corresponding to the offending symbols (that is, the larger of each inconsistent pair). If so desired, we can now rename our symbols to form a set of consecutive natural numbers beginning with 1. The interrupted scanning process can now continue, and we are back in the old routine. The situation is well illustrated by the following example.

Example 3. We apply the method to the group

$$T_2 = \langle x,y \mid x^2y^3,\ x^3y^4 \rangle,$$

and continue until a collapse occurs (at $\equiv$), whereupon the partially completed tables are as follows:

	x		x		y		y		y	
1	–	2	–	3	–	4	–	5	=	1
2		3		6		3		4	$\equiv$	2
3										3
4								3		4
5						3		4		5
6										6

	x	x		x		y	y	y		y	
1	2	3	–	6	=	3	4	5			1
2	3										2
3											3
4								3			4
5							3	4			5
6											6

Table 8

definition	bonus		x	y
1x = 2		1	2	
2x = 3		2	3	
3y = 4		3	6	4
*... 4y = 5	5y = 1	4		5
3x = 6	6y = 3	5		1
	4y = 2 ...*	6		3

Table 9

The inconsistent information (labelled *) tells us that $5 = 2$. We replace 5 by 2 and 6 by 5 throughout and continue scanning. As there are no further collapses, our relator tables now appear as follows:

	x	x	y	y	y
1	2	3	4	2	1
2	3	5	3	4	2
3	5			5	3
4			5	3	4
5					5

	x	x	x	y	y	y	y
1	2	3	5	3	4	2	1
2	3	5		5	3	4	2
3	5					5	3
4				4	3		4
5							5

Table 10

Defining $5x = 6$, the second row of the second table gives $6y = 5$, and completion of the third row of the first table gives the collapse $6 = 5$. Hence $5y = 5$ and together with the equations

$$5y = 3, \ 3y = 4, \ 4y = 2, \ 2y = 1 \ ,$$

this leads to successive collapses $5 = 3 = 4 = 2 = 1$. *Since the first row of both tables is complete*, we are not surprised to deduce that T_2 is just the trivial group.

This example highlights the value of the strategy of defining new symbols in such a way that the first rows of all the relator tables become complete as quickly as possible: this ploy was adopted in both of the earlier examples. Another point worthy of note is provided by the following generalization.

Example 4. It is plain that the group

$$T_n = <x,y \mid x^n y^{n+1}, x^{n+1} y^{n+2}>, \ n \in \mathbb{N} \ ,$$

is trivial, and that the first definition must be of the form $1x = 2$, or $1y^{-1} = 2$. In fact the definitions must continue to be of the form $ix = j$ or $iy^{-1} = j$ (with $i < j$) for some time. It is not hard to see that at least the first n definitions must be of this type (see Exercise 2), and that no row can become complete before the nth definition is made. Thus, for any $n \in N$, there is a presentation of the trivial group for which the coset enumeration requires the use of more than n symbols. While the process will always terminate for finite groups, there is in general no way of bounding in advance the number of symbols that will be needed. For this reason, the method cannot be described as an algorithm.

2.   A refinement

By means of a very simple adjustment, the method described in the preceding section can be carried out relative to a subgroup H of a finitely presented group $G = <X \mid R>$. H is specified as the subgroup of G generated by a finite set Y of words in $X^{\pm 1}$, and the process terminates after a finite number of steps provided that $|G : H|$ is finite. We obtain such information as the

>   index of H in G,

>   the permutation representation of G on the (right) cosets of H,

>   the coset diagram of G relative to H, and

>   a transversal for H in G with the Schreier property.

While it yields less information than the old method in general, the advantage of the new method is that the tables become complete much sooner. When $H = E$, the two methods yield the same information.

At the outset we draw up relator and monitor tables as before, and in addition a table for each generator y of H. These new tables are contructed in the same way as the relator tables, with the letters of y separating adjacent columns, except that they have only one row, beginning and ending with the symbol 1. The method then proceeds exactly as above, with the Y-tables being completed according to the same rules as the R-tables. The process again terminates when there are no more empty spaces, whereupon $|G : H|$ is just the number of rows in each R-table.

Example 5. We begin with a simple example, to which the answer is already known. Letting $G = <x \mid x^6>$ and $H = <x^3> \leq G$, the tables are as follows:

x	x	x	x	x	x	
1	2	3	1	2	3 $\equiv$	1
2	3	1	2	3 $\equiv$	1	2
3	1	2	3 $\equiv$	1	2	3

x	x	x	
1 – 2 – 3 $=$	1		

definition	bonus
$1x = 2$	
$2x = 3$	$3x = 1$

	x
1	2
2	3
3	1

Table 11

The one-rowed table is the first to complete, and the resulting bonus is sufficient for the three rows of the relator table to complete without yielding new information. We deduce that $|G : H| = 3$ and obtain the permutation representation $x \mapsto (123) \in S_3$ for G. The Schreier transversal for H in G is just $\{e, x, x^2\}$, and the coset diagram is as follows:

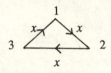

Fig. 14

Example 6. We carry out the same process for the von Dyck group $D(3,3,2)$:

$$G = \langle x, y \mid x^3, y^3 (xy)^2 \rangle \ ,$$

with respect to the subgroup $H = \langle x \rangle$. The tables are as follows:

	x		x		x		
1	1		1	≡	1		
2	3	–	4	=	2		
3	4	≡	2		3		
4	≡	2		3	4		

	y		y		y		
1	–	2	–	3	=	1	
2	3	≡	1		2		
3	≡	1		2	3		
4	4		4	≡	4		

	x		y		x		y		
1	1		2	=	3		1		
2	3	≡	1		1		2		
3	4	=	4		2		3		
4	2		3		4	≡	4		

	x		
1	=	1	

definition	bonus
	$1x = 1$
$1y = 2$	
$2y = 3$	$3y = 1$
	$2x = 3$
$3x = 4$	$4x = 2$
	$4y = 4$

	x	y
1	1	2
2	3	3
3	4	1
4	2	4

Table 12

We deduce that $|G : H| = 4$, and that $\{e, y, y^2, y^2x\}$ is a Schreier transversal. We also have the permutation representation

$$\alpha: \; G \to S_4$$
$$x \mapsto (234)$$
$$y \mapsto (123)$$

and we deduce that $|x| \geq |x\alpha| = 3$. Since x^3 is a relator, we already know that $|x| \leq 3$, and conclude that $|H| = |x| = 3$. It follows that $|G| = 12$, and since $A_4 = <(234), (123)> = \text{Im } \alpha$, we have identified G up to isomorphism.

This example illustrates a number of important points of which some are typical and others not. Firstly, the situation where the generators of H form a subset of the generators

of G (that is, $Y \subseteq X$) is a particularly propitious one. In such cases, we often omit the one-rowed tables altogether and simply subsume the corresponding information into the second column of the first monitor table. Next, the periodic nature of the relators results in some redundancies in the relator tables. When this happens, it saves time to replace the internal entries of a superfluous row by a dash linking the first and last entries as in Example 7 below. Finally, the fact that G has been identified, that is, α is faithful, is by no means typical (cf. Example 5 and Exercise 14) and must be regarded simply as a stroke of good luck. The next example illustrates each of these points, and we counsel the reader to carry out the computation for himself.

Example 7. We enumerate cosets with respect to $H = <x>$ in $D\,(4,3,2)$:

$$G = <x,y\,|\,x^4,y^3,(xy)^2> .$$

	x	x	x	x	
1	1	1	1	1
2	3 — 4 — 5			$=$	2
3	———————————————				3
4	———————————————				4
5	———————————————				5
6	6	6	6	$\equiv$	6

	y	y	y				x	y	x	y
1	— 2 — 3		$=$	1		1	1	2 $=$ 3		1
2	———————			2		2	———————————			2
3	———————			3		2	4 $=$ 5		2	4
4	5 — 6		$=$	4		4	5	6 $=$ 6		4
5	———————			5		5	———————————			5
6	———————			6		6	———————————			6

Table 13

definition	bonus
	$1x = 1$
$1y = 2$	
$2y = 3$	$3y = 1$ $2x = 3$
$3x = 4$	
$4x = 5$	$5x = 2$ $4y = 5$
$5y = 6$	$6y = 4$ $6x = 6$

	x	y
1	1	2
2	3	3
3	4	1
4	5	5
5	2	6
6	6	4

Table 14

By the same argument as used in the previous example, we deduce that $|G| = 24$, and it is not hard to show (Exercise 8) that $G \cong S_4$. We leave it to the reader to work out the other by-products of the computation.

Example 8. To illustrate the phenomenon of coset collapse in the adapted method, we consider the group

$$F(2,5) = <x,a,b,c,d \mid xa = b,\ ab = c,\ bc = d,\ cd = x,\ dx = a> ,$$

and enumerate cosets with respect to $<x>$.

x	a			b^{-1}
1	1 – 3	=		1
2	3			2
3	2			3

	a		b		c^{-1}
1	3	=	2		1
2	1		3	=	2
3					3

	b		c		d^{-1}
1	3	≡	2		1
2			1		2
3	2		3	=	3

	c		d		x^{-1}
1	– 2	=	1		1
2	3		3	=	2
3					3

	d		x		a^{-1}
1	= 2		3		1
2	3		1	=	2
3	3		3		3

definition	bonus		x	a	b	c	d
	$1x = 1$	1	1	3	3	2	2
$\underline{1c} = 2$	$2d = 1$ $2a = 1$	2	3	1		3	1
$1a = 3$	$1b = 3$ $3b = 2$ $2c = 3$ $3d = 3$	3			2		3
	$2x = 3$ $1d = 2$ $\underline{3c = 2}$						

Table 15

At this point we deduce that $1 = 2c^{-1} = 3$, whence $3 = 1b = 3b = 2$, and the tables collapse to one row each. The second monitor table tells us that each of the five generators fixes 1, so that $F(2,5) = <x>$ and is thus abelian. Since we already know (Exercise 6.6(iii)) that the derived factor group of $F(2,5)$ is Z_{11}, we deduce that $F(2,5) \cong Z_{11}$ (as in the table of Chapter 7).

Exercises

1. Use the permutation representation given by the enumeration in Example 2 to identify $F(2,3)$. Can the corresponding Cayley diagram be embedded in the plane (or 2-sphere, or torus)?

2. Prove in detail that in enumerating cosets for the presentation

    $$T_n = <x,y \mid x^n y^{n+1}, \; x^{n+1} y^{n+2}>, \; n \in \mathbb{N} \; ,$$

    at least $n + 1$ symbols are needed. Can you enlarge this lower bound?

3. Let $G = <X \mid R>$ be a finitely presented group, and n a natural number. Find a presentation for G for which the coset enumeration requires more than n symbols.

4. Prove that the words $u_1, ..., u_g$ defined after Table 5 always have the Schreier property.

5. Let F be the free group on $\{a,b\}$ and $R = \{baba^{-1}, \; abab^{-1}\} \subseteq F$. Use the first monitor table in Example 2 to write down a Schreier transversal for $\bar{R}$ in F.

6. Draw a coset diagram for the enumeration carried out in Example 6.

7. Describe the Schreier transversal and permutation representation resulting from Example 7, and draw the coset diagram.

8. In S_4, find a 4-cycle and a 3-cycle which generate the group and whose product has order 2. Deduce from the result of Example 7 that $D(4,3,2) \cong S_4$.

9. Prove that $D(5,3,2) \cong A_5$.

10. Enumerate the cosets of the subgroup $H = <xy, zxyz>$ in the group

    $$G = <x,y,z \mid x^2, y^2, z^2 (xy)^3, (yz)^3, [x,z]> \; .$$

11. Identify the group

    $$<x,a,b,c,d \mid x = bd, \; a = cx, \; b = da, \; c = xb, \; d = ac> \; .$$

12. Compute the order of the group

    $$<x,y \mid x^4 = y^4 = e, \; yx = x^2 y^2> \; .$$

13. Find the index of $<x>$ in the group

    $$G = <x,y \mid x^2 yxy^3, y^2 xyx^3> \; .$$

14. Let ρ be the permutation representation of a group G obtained by enumerating cosets with respect to a subgroup H. Prove the Ker ρ is equal to the intersection of the conjugates of H. (This is the largest normal subgroup of G lying in H, and is called the *core* of H.)

15. Given a complete enumeration of the cosets of $H = <Y>$ in G, prove that H is normal in G if and only if every $y \in Y$ fixes every symbol.

CHAPTER 9

PRESENTATIONS OF SUBGROUPS

Whereas presentations of factor groups (Proposition 4.2) and group extensions (Proposition 10.1) are relatively easy to come by, finding generators and defining relations for a subgroup H of a group $G = <X \mid R>$ is less straightforward. It is achieved, however, by means of the celebrated Reidemeister-Schreier rewriting process, which is arguably the most pervasive method in the armoury of the combinatorial group theorist.

1.   The method

For practical purposes, the method involves four steps, which yield in turn the following information:

(i)   a (Schreier) transversal U for H in G,

(ii)   generators B of H,

(iii)   defining relators $\hat{R} = \hat{R}(X)$ for H,

(iv)   defining relators $\hat{S} = \hat{S}(B)$ for H.

The implementation of Step (i) depends on the manner in which H is specified. For example, we may be told that H is the subgroup generated by a given set of words $Y = Y(X)$. Then the method of §8.2 yields U, provided $|G : H| < \infty$. On the other hand, H may be some normal subgroup of G such that the structure of the factor group G/H is known; when $H = G'$ for example, the methods of §6 apply.

The generators B in Step (ii) are just the $ux\overline{ux}^{-1}$ of §2.4 ($u \in U$, $x \in X$, $ux \notin U$). In order to write them down we need to know not only X and U, but also the function $F(X) \to U$ that assigns to any word w in $X^{\pm}$, regarded as an element of G, its coset representative $\overline{w} \in U$. This information is forthcoming in both the cases mentioned above (the methods of §§8.2 and 6.4).

Step (iii) is purely mechanical and will be described presently, it consists in merely writing down the words in $\hat{R} = \{uru^{-1} \mid u \in U, r \in R\}$.

The last step is the rewriting process whereby each element of $\hat{R}$ is expressed in terms of the generators B. This may be achieved by the simple algorithm of Lemma 2.3, although it is usually more fun to do it by inspection.

The resulting presentation $H = <B \,|\, \hat{S}>$ is often rather unwieldly: if

$$|X| = r, \quad |R| = m, \quad |G:H| = g \;,$$

then

$$|B| = (r-1)g + 1, \quad |\hat{S}| = mg \;.$$

The usual course is then to reduce it at leisure to recognizable form by judicious use of Tietze transformations (§4.4).

To overall process is thus a happy combination of almost every technique discussed in the foregoing. The only missing ingredient is the rather trifling one in Step (iii) and we supply this now.

Lemma 1. Given a subgroup K of a group F, let R be a subset of F whose normal closure $\bar{R}$ in F lies inside K. Then $\bar{R}$ is the normal closure in K of the set

$$\hat{R} = \{uru^{-1} \,|\, u \in U, \; r \in R\} \;, \tag{1}$$

where U is any transversal for K in F.

Proof. $\bar{R}$ is generated by the elements wrw^{-1}, $w \in F$, $r \in R$. But any $w \in F$ can be written in the form ku, $k \in K$, $u \in U$. Thus, $\bar{R}$ is generated by the elements $kuru^{-1} k^{-1}$, $k \in K$, $u \in U$, $r \in R$, so that $\hat{R}$ is generated by the conjugates in K of $\hat{R}$, as required.

It is now easy to obtain defining relators for our subgroup H of G. Merely let

$$F = F(X), \quad v = \text{nat} : F(X) \to G, \quad K = v^{-1}(H) \;.$$

Then, by abuse of notation, U is a Schreier transversal for K in F and the set B freely generates K (Theorem 2.1). It now follows from Lemma 1 that $\bar{R}$ is the normal closure in K of the set $\hat{R}$ defined by (1). Letting $\hat{S}$ denote the set of elements of $\hat{R}$ written as words in $B^{\pm}$, $<B \,|\, \hat{S}>$ is a presentation for the group $K/\bar{R}$, which is isomorphic to H.

Proposition 1. Let $H \leq G = <X \,|\, R>$. Then with the above notation, $H = <B \,|\, \hat{S}>$.

Corollary 1. Let H be a subgroup of finite index in a group G. If G is finitely presented, then so is H.

The remainder of this chapter is given over to examples. In so far as they serve to illustrate the method, all are of practical interest. Many also boast theoretical importance.

2. Alternating groups

According to formula (5.5), the symmetric group S_4 has the presentation $<X\,|\,R>$, where

$$X = \{x_1, x_2, x_3\}$$

$$R = \{x_2^1, x_2^2, x_3^2, (x_1 x_2)^3, (x_2 x_3)^3, x_1^{-1} x_3^{-1} x_1 x_3\}. \tag{2}$$

Since x_i stands for the transposition $(i\ i+1)$, $1 \le i \le 3$, we may take $U = \{e, x_1\}$ as a transversal for A_4 in S_4. By the parity rules for multiplying permutations, it is clear that $\overline{ex_i} = x$, and $\overline{x_1 x_i} = e$, $1 \le i \le 3$. This yields the left-hand half of the following table (cf. Table 21).

	x_1	x_2	x_3	x_1^2	x_2^2	x_3^2	$(x_1 x_2)^3$	$(x_2 x_3)^3$	$x_1^{-1} x_3^{-1} x_1 x_3$
e	$-$	$x_2 x_1^{-1}$	$x_3 x_1^{-1}$	x_1^2	x_2^2	x_3^2	$(x_1 x_2)^3$	$(x_2 x_3)^3$	$x_1^{-1} x_3^{-1} x_1 x_3$
x_1	x_1^2	$x_1 x_2$	$x_1 x_3$	x_1^2	$x_1 x_2^2 x_1^{-1}$	$x_1 x_2^3 x_1^{-1}$	$x_1(x_1 x_2)^3 x_1^{-1}$	$x_1(x_2 x_3)^3 x_1^{-1}$	$x_3^{-1} x_1 x_3 x_1^{-1}$

Table 16

The rows in both havles of the table are indexed by U, and the columns by X, R, respectively. The (u,x)-entry is $ux\overline{ux}^{-1}$ and the (u,r)-entry is uru^{-1}. We call this a $B\hat{R}$-table. The corresponding $B\hat{S}$-table is obtained from it by relabelling the generators B by new symbols and rewriting the relators $\hat{R}$ in terms of these. To express $h = x_3^{-1} x_1 x_3 x_1^{-1}$, for example, in terms of the new generators, the inductive process in the proof Lemma 2.3 goes like this:

$$u_1 = e, \qquad u_2 = \overline{ex_3^{-1}} = x_1, \qquad a_1 = ex_3^{-1} x_1^{-1},$$

$$u_2 = x_1, \qquad u_3 = \overline{x_1 x_1} = e, \qquad a_2 = x_1 x_1,$$

$$u_3 = e, \qquad u_4 = \overline{ex_3} = x_1, \qquad a_3 = ex_3 x_1^{-1},$$

$$u_4 = x_1, \qquad u_5 = \overline{x_1 x_1^{-1}} = e, \qquad a_4 = x_1 x_1^{-1} = e,$$

so that $h = a_1 a_2 a_3$, with $a_j \in B^{\pm}$, $1 \le j \le 3$. This gives the last entry in the $B\hat{S}$-table below, the other entries in the right-hand half being bound by inspection.

$x_1\ x_2\ x_3$

| $-$ | $b_2\ b_4$ | b_1 | b_2b_3 | b_4b_5 | b_3^3 | $(b_2b_5)^3$ | $b_1^{-1}b_4^{-1}b_5$ |

| b_1 | $b_3\ b_5$ | $-$ | b_3b_2 | b_5b_4 | $(b_1b_2)^3$ | $(b_3b_4)^3$ | $b_5^{-1}b_1b_4$ |

Table 17

Here and in subsequent examples, the row-headings r are omitted, since they coincide with the entries ere^{-1} directly beneath them. The repeated relator $x_1^{-1}x_1^2x_1 = x_1^2 = b_1$ is also left out.

To reduce this to manageable form, make informal use of Tietze transformations as follows. First delete the trivial generator b_1, then use the relators b_2b_3, b_4b_5 to replace b_2, b_5 by b_3^{-1}, b_4^{-1}, respectively. The only remaining non-trivial relations are those in the last three columns of Table 2, namely,

$$b_3^3,\ (b_3^{-1}b_4^{-1})^3,\ b_4^{-2},\ b_3^{-3},\ (b_3b_4)^3,\ b_4^2\ .$$

Since the first three are consequences of the last three, we find that

$$A_4 = <b_3, b_4 \,|\, b_3^{-3}, b_4^2, (b_3b_4)^3 >$$

$$= D\,(-3,2,3) = D\,(3,3,2)\ ,$$

by Exercise 4.3. This confirms the result of Example 8.6.

3. Braid groups

Consider the defining relators (2) for S_4. Since x_1 and x_2 have order 2, the fourth relator can be written, as a relation, in the form

$$x_1x_2x_1 = x_2x_1x_2\ .$$

This is a *braid relation*, which we abbreviate to $x_1 \approx x_2$. Writing $x_1 \sim x_3$ to denote that x_1 commutes with x_3, we may define the Artin *braid group* B_n on n strings as follows:

$$B_n = <x_1,x_2,...,x_{n-1}\,|\,x_1 \approx x_2 \approx ... \approx x_{n-1},\ \{x_i \sim x_j \,|\, |i-j| \geq 2\}>\ . \qquad (3)$$

Thus, "adjacent" generators stand in the braid relation to one another, while non-adjacent generators commute. Throwing in the extra relations $x_i^2 = e$, $1 \leq i \leq n-1$, yields the presentation (5.5) of S_n. It follows from von Dyck's theorem (Proposition 4.2) that S_n is a homomorphic image of B_n. Let $v : B_n \twoheadrightarrow S_n$ denote the corresponding natural map.

The groups B_n have far-reaching geometrical significance. Their structure is arrived at by exploiting the epimorphism $v : B_n \twoheadrightarrow S_n$ of the previous paragraph to obtain a suitable subgroup H as input for the Reidemeister-Schreier process. We shall carry this out in the case $n = 3$, which is fairly typical

Specifically, take

$$B_3 = <x_1, x_2 \mid x_1 x_2 x_1 x_2^{-1} x_1^{-1} x_2^{-1} >$$

(cf. Example 4.6) and let H be the pre-image under v of the stabilizer S of 1 in S_3. Thus, $S \cong S_2 = \text{Sym}(\{2,3\})$, and $|B_3 : H| = |S_3 : S| = 3$. We may take

$$U = \{e, x_1, x_1 x_2\}$$

as these four elements map 1 respectively to 1,2,3. The values of the "bar" function are found by applying v and working in S_3 where, by abuse of notation, $x_i = (i\ i + 1)$, $i = 1,2$. Then the $B\hat{R}$-table is as follows, where, for example, the $(x_1 x_2, x_1)$-entry is arrived at

U	x_1	x_2	
e	$-$	x_2	$x_1 x_2 x_1 x_2^{-1} x_1^{-1} x_2^{-1}$
x_1	x_1^2	$-$	$x_1^2 x_2 x_1 x_2^{-1} x_1^{-1} x_2^{-1} x_1^{-1}$
$x_1 x_2$	$x_1 x_2 x_1 x_2^{-1} x_1^{-1}$	$x_1 x_2^2 x_1^{-1}$	$x_1 x_2 x_1 x_2 x_1 x_2^{-1} x_1^{-1} x_2^{-2} x_1^{-1}$

Table 18

as follows: $x_1 x_2 x_1$ passes under v to $(12)(23)(12) = (13) \in S_3$, and thus lies in the same coset of S as $x_1 x_2 = (12)(23) = (132)$, since both send 1 to 3. We obtain the $B\hat{S}$-table by inspection, where a fairly good rule of thumb is that the rewritten (u,r)-entry begins with a new generator in the same row when $u \neq e$ (cf. Table 17).

	$-$	b_3	$b_2 b_3^{-1}$
b_1		$-$	$b_1 b_3 b_4^{-1} b_2^{-1}$
b_2		b_4	$b_2 b_4 b_1 b_3^{-1} b_1^{-1} b_4^{-1}$

Table 19

Eliminating $b_3 = b_2$ by a Tietze transformation, it follows that H is generated by b_1, b_2, b_4, with defining relations

$$b_1 b_2 b_4^{-1} b_2^{-1} = b_2 b_4 b_1 b_2^{-1} b_1^{-1} b_4^{-1} = e \ ,$$

or equivalently,

$$b_4^{b_2^{-1}} = b_1, \quad (b_4 b_1)^{b_2^{-1}} = b_4 b_1 \ .$$

Relabelling b_4, b_1 as a_1, a_2 and b_2^{-1} as x_1, we have the presentation

$$H = <a_1, a_2, x_1 \mid a_1^{x_1} = a_2, (a_1 a_2)^{x_1} = a_1 a_2> \ . \tag{4}$$

It is clear from the form of (4) that the subgroup A generated by a_1, a_2 is normal, and that A and $B = <x>$ together generate H. It will become clear in the next chapter that A and B are actually free on the given generators, and that H is their semi-direct product $A \,] \, B$ (see §5.4) with respect to the homomorphism $\gamma : B \to \text{Aut } A$ given by

$$a_1(x_1 \, \gamma) = a_2, \quad a_2(x_1 \, \gamma) = a_2^{-1} a_1 a_2 \ . \tag{5}$$

We end this section by giving the result of the corresponding computation in the general case. Let H be the pre-image under $v : B_n \to S_n$ of Stab $(1) \leq S_n$, so that $\mid B_n : H \mid = n$. Then it turns out that H is the semi-direct product of a free group of rank $n - 1$ with the braid group B_{n-1} on $n - 1$ strings with respect to an action that suitably generalizes (5). This provides the key to an inductive process whereby many questions about the structure of B_n can be answered.

4.   von Dyck groups

The groups

$$D \, (l, m, n) = <x, y \mid x^l, y^m, (xy)^n>$$

of Example 4.5 are really geometrical objects, and as such have been a favoured object of study since antiquity. We shall give a brief account of their algebraic structure, making only passing reference to the geometrical ideas that underlie it.

By Exercise 4.3, we can assume without loss of generality that $l \geq m \geq n \geq 1$. The degenerate cases $n = 1$ and $m = n = 2$ are covered in Example 4.4, yielding cyclic and dihedral groups, respectively. The cases studied in

Example 8.6, Example 8.7, Exercise 8.8

are given by

$$(l, m, n) = (3, 3, 2), \ (4, 3, 2), \ (5, 3, 2) \ .$$

and the corresponding groups are

$$A_4, \quad S_4, \quad A_5 \ ,$$

which are just the (orientation-preserving) symmetry groups of the

tetrahedron, octahedron, icosahedron,

respectively. It is easy to see (Exercise 5) that this covers all positive solutions of the Diophantine inequality

$$1/l + 1/m + 1/n > 1 \ ,$$

which we refer to as the *spherical case*.

In the *Euclidean case*

$$1/l + 1/m + 1/n = 1 \ ,$$

there are only three solutions, namely

$$(l,m,n) = (3,3,3), \ (4,4,2), \ (6,3,2)$$

(see Exercise 5), so that the vast majority of the $D\ (l,m,n)$ belong to the *hyperbolic case*

$$1/l + 1/m + 1/n < 1 \ .$$

These three cases correspond to geometries in which the angle-sum of a triangle is greater than, equal to, or less than π, respectively. We shall apply the Reidemeister-Schreier rewriting process to one example in each case.

Example S: (3,3,2).

We reconfirm the result of Example 8.6 by finding a presentation for the subgroup $H = <x>$ in the group

$$G = <x,y \mid x^3, y^3 (xy)^2> \ .$$

Our chief aim here is to show how a Schreier transversal and the values of the bar function may be read off from the first and second columns of the "definition-bonus" table, respectively, which we repeat here for convenience. Taking $1 = H = He$,

definition	bonus
	$1x = 1$
$1y = 2$	
$2y = 3$	$3y = 1$
	$2x = 3$
$3x = 4$	$4x = 2$
	$4y = 4$

Table 20

the first column gives

$$2 = 1y = Hy, \quad 3 = 2y = 1y^2 = Hy^2, \quad 4 = 3x = 2yx = 1y^2x = Hy^2x \ ,$$

so that the cosets 1,2,3,4 are represented by

$$e, \ y, \ y^2, \ y^2x \ ,$$

respectively. Taking this as U, the five equations in the second column yield

$$\overline{ex} = e, \ \overline{y^3} = e, \ \overline{yx} = y^2, \ \overline{y^2x^2} = y, \ \overline{y^2xy} = y^2x \ ,$$

respectively. We are now in a position to compile the $B\hat{R}$-table (Table 21). To save space in the $B\hat{S}$-table (Table 22), we replace each b_i by its subscript i, and write $\overline{i}$

U	x	y			
e	x	$-$	x^3	y^3	$(xy)^2$
y	yxy^{-2}	$-$	yx^3y^{-1}	$-$	$(yx)^2$
y^2	$-$	y^3	$y^2x^3y^{-2}$	$-$	y^2xyxy^{-1}
y^2x	$y^2x^2y^{-1}$	$y^2xyx^{-1}y^{-2}$	$-$	$y^2xy^3x^{-1}y^{-2}$	$y^2x^2yxyx^{-1}y^{-2}$

Table 21

rather than i^{-1} for b_i^{-1}. Since $3 = e = 12 = 24 = 01$, H is the cyclic group generated by

```
0 –   0³  3  013
1 –   12  –  130
– 3   21  –   42
2 4   –   4³  24
```

Table 22

$0(= \bar{1} = 2 = \bar{4})$ and defined by 0^3, that is, $H \cong Z_3$, as expected.

Example E: (3,3,3).

We find a presentation for $H = G'$ in the group

$$G = \langle x,y \mid x^3 = y^3 = (xy)^3 = e \rangle \ .$$

Using the results of Chapter 6, we have $G_{ab} = Z_3 \times Z_3$, and we can take $U = \{x^i y^i \mid 0 \le i,j \le 2\}$. The bar function is calculated by working in G_{ab}, that is, from the rules

$$\left. \begin{array}{l} x^i y^j x \equiv x^{i+1} y^j \equiv x^{i-2} y^j \\ x^i y^j y = x^i y^{j+1} \equiv x^i y^{j-2} \end{array} \right\} \quad (\text{mod } G') \ .$$

The $B\hat{R}$-table is a shown in Table 21, and the $B\hat{S}$-table in Table 24 (where the b's are again suppressed).

U	x	y			
e	–	–	x^3	y^3	$(xy)^3$
x	–	–	–	xy^3x^{-1}	$x^2yxyxyx^{-1}$
x^2	x^3	–	–	$x^2y^3x^{-2}$	$x^3yxyxyx^{-2}$
y	$yxy^{-1}x^{-1}$	–	yx^3y^{-1}	–	$(yx)^3$
xy	$xyxy^{-1}x^{-2}$	–	$xyx^3y^{-1}x^{-1}$	–	–
x^2y	x^2yxy^{-1}	–	$x^2yx^3y^{-1}x^{-2}$	–	–
y^2	$y^2xy^{-2}x^{-1}$	y^3	$y^2x^3y^{-2}$	–	y^2xyxyx^{-1}
xy^2	$xy^2xy^{-2}x^{-2}$	xy^3x^{-1}	$xy^2x^3y^{-2}x^{-1}$	–	$xy^2xyxyxy^{-1}x^{-1}$
x^2y^2	$x^2y^2xy^{-2}$	$x^2y^3x^{-2}$	$x^2y^2x^3y^{-2}x^{-2}$	–	$x^2y^2xyxyxy^{-1}x^{-2}$

Table 23

```
                    - -    0  7  267
                    - -    -  8  348
                    0 -    -  9  0159
                    1 -  123  - 1590
                    2 -  231  -   -
                    3 -  312  -   -
                    4 7  456  - 483
                    5 8  564  - 5901
                    6 9  645  - 672
```

Table 24

Deleting the trivial generators 0, 7, 8, 9 and cyclic conjugates of known relators, we obtain the presentation

$$H = \langle 1,2,3,4,5,6 \mid 123, 456, 26, 34, 15 \rangle$$

$$= \langle 1,2,3 \mid 123, \overline{\overline{312}} \rangle$$

$$= \langle 1,2 \mid 12\overline{\overline{12}} \rangle \ ,$$

which is clearly the free abelian group of rank 2. Thus, $H \cong Z \times Z$. The same result is obtained in the other two Eulidean cases. That the rank is 2 in every case is a reflection of the fact that the underlying space is 2-dimensional ($\mathbb{R}^2$ in this case).

Example H: (3,3,4).

It follows from the last relation in

$$G = \langle x,y \mid x^3 = y^3 = (xy)^4 = e \rangle$$

that x and y^{-1} lie in the same coset of $H = G'$. Thus, $|G : H| = 3$ and we may take e, x, x^2 as the transversal. The tables are then as follows.

	x	y			
e	$-$	yx^{-2}	x^3	y^3	$(xy)^4$
x	$-$	xy	$-$	xy^3x^{-1}	$x(xy)^4x^{-1}$
x^2	x^3	x^2yx^{-1}	$-$	$x^2y^3x^{-2}$	$x^2(xy)^4x^{-2}$

Table 25

$$
\begin{array}{lll}
-\,a & d\ acb & b^4 \\
-\,b & -\,bac & c^4 \\
d\ c & -\,cba & (da)^4
\end{array}
$$

Table 26

This yields

$$H = <a,b \mid a^4 = b^4 = (ba)^{-4} = e> \; ,$$

so that $D\,(3,3,4)' \cong D\,(4,4,4)$.

The same process applied to $D\,(4,4,4)$ yields a 6-generator 1-relator presentation for its derived group. The relator is a word of length 12 in which each generator and its inverse appear exactly once. Suitable Tietze transformations then yield the presentation

$$D\,(4,4,4)' = <a,a',b,b',c,c' \mid [a,a']\,[b,b']\,[c,c']> \; . \tag{6}$$

If we label and orient the edges of dodecagon as in Fig. 15 (this diagram is to be thought of as a 2-cell: cf. the discussion in §4.5)

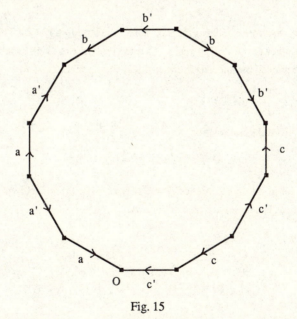

Fig. 15

the boundary-label, read counter-clockwise from the point O, is just the relator (6). By pasting together corresponding pairs of sides in the sense indicated by the arrows, we

obtain a surface S homeomorphic to a sphere with three handles. Then (6) is a presentation for its fundamental group $\pi_1(S)$.

5. Triangle groups

If l_1 and l_2 are two lines that meet in a point O (see Fig. 16) and form an angle θ, then the result of composing a reflection in l_1 with a reflection in l_2 is just a rotation about O through 2θ. Now let Δ be a triangle with angles π/l, π/m, π/n, where l, m, n are integers greater than 1, and let a,b,c denote reflections in the three sides of Δ (see Fig. 17). The corresponding *triangle group* is then given by

$$\Delta(l,m,n) = <a,b,c \mid a^2 = b^2 = c^2 = (ab)^l = (bc)^n = (ca)^m = e> . \tag{7}$$

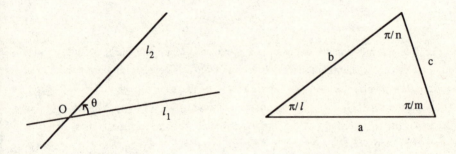

Fig. 16   Fig. 17

Since every relator in (7) has even length, there is a homomorphism from $G = \Delta(l,m,n)$ onto $Z_2 = <t \mid t^2>$ sending every generator to t. Its kernel H consists of all words of even length. Taking $U = \{e,a\}$, we obtain Table 12 and 13.

	a	b	c						
e	$-$	ba^{-1}	ca^{-1}	a^2	b^2	c^2	$(ab)^l$	$(bc)^m$	$(ca)^n$
a	a^2	ab	ac	$-$	ab^2a^{-1}	ac^2a^{-1}	$a(ab)^la^{-1}$	$a(bc)^ma^{-1}$	$(ac)^n$

Table 27

$$-2\ 4 \quad 1\ 23\ 45 \quad 3^l \quad (25)^n \quad (41)^m$$
$$1\ 3\ 5 \quad -32\ 54 \quad (12)^l \quad (34)^n \quad 5^m$$

Table 28

Taking generators $x = 3 = ab$, $y = 4 = ca^{-1}$, we see that

$$H = <x,y \,|\, x^l = y^m = (xy)^n = e> \ .$$

$D\,(l,m,n)$ is thus a subgroup of index 2 in $\Delta\,(l,m,n)$. Its elements are just the transformations that preserve orientation.

6.   Free products

Given groups $G = <X\,|\,R>$ and $H = <Y\,|\,S>$, their *free product* is given by the presentation

$$G*H = <X,Y\,|\,R,S> \ .   \tag{8}$$

The free product can be defined by a universal property (a pushout) dual to that defining the direct product (a pullback). It follows that $G*H$ depends only on G and H, and not on the presentations used in (8). The kernel C of the natural map: $G*H \rightarrow G \times H$ is called the *Cartesian* of $G*H$. It is a consequence of the Kurosh subgroup theorem (which is the analogue for free products of the Nielsen-Schreier theorem for free groups) that C is always a free group. We illustrate this by an example.

Let

$$G = <x,y \,|\, x^2, y^3> \cong Z_2 * Z_3 \ ,$$

and let H be the Cartesian. Since the free factors are abelian, H coincides with G', and we can take $U = \{x^i y^j \,|\, 0 \le i \le 1,\ 0 \le j \le 2\}$. The tables are as follows.

	x	y	x^2	y^3
e	–	–		
x	x^2	–		xy^3x^{-1}
y	$yxy^{-1}x^{-1}$	–	yx^2y^{-1}	–
xy	$xyxy^{-1}$	–	$xyx^2y^{-1}x^{-1}$	–
y^2	$y^2xy^{-2}x^{-1}$	y^3	$y^2x^2y^{-2}$	–
xy^2	xy^2xy^{-2}	xy^3x^{-1}	$xy^2x^2y^{-2}x^{-1}$	–

Table 29

$$
\begin{array}{cccc}
- & - & 1 & 6 \\
1 & - & - & 7 \\
2 & - & 23 & - \\
3 & - & 32 & - \\
4 & 6 & 45 & - \\
5 & 7 & 54 & -
\end{array}
$$

Table 30

It is clear that H is free of rank 2, with free generators

$$2 = yxy^{-1}x^{-1} \quad \text{and} \quad 4 = y^2xy^{-2}x^{-1} \ .$$

A more general construction is the *free product with amalgamation*. As before, let $G = <X \mid R>$, $H = <Y \mid S>$. Suppose further that G has a subgroup $K = <Z>$ and that $\phi : K \to H$ is a monomorphism. Then we define

$$G*_KH = <X,Y \mid R,S, \{(z\phi)\,z^{-1} \mid z \in Z\}> \ . \tag{9}$$

The braid group B_3 is of this type: let

$$G = <a \mid>, \ H = <b \mid>, \ Z = \{a^3\}, \ a^3\phi = b^2 \ ,$$

so that G,H,K are all infinite cyclic. The presentation (9) is then

$$G*_KH = <a,b \mid a^3 = b^2> \ ,$$

which is isomorphic to B_3 (see Example 4.6).

7.   HNN-extensions

Now let $G = <X \mid R>$, $K = <Z> \leq G$ and $\phi : K \to G$ a monomorphism. Then the corresponding HNN-extension is given by

$$G*_K\phi = <X,t \mid R = e, \ \{z\phi = t^{-1}zt \mid z \in Z\}> \ . \tag{10}$$

This group depends only on G and the *associated subgroups* K and $K\phi$. t is called the *stable letter*. There is a normal theorem for HNN-extensions from which it follows that the natural mapping $G \to G*_K\phi$ is an embedding. We give three examples.

First, let $F = <x,y \mid>$ denote the free group of rank 2, and consider the mapping

$$\phi: \quad \left.\begin{array}{l} F \to F \\ x \mapsto x^2 \\ y \mapsto y^2 \end{array}\right\} \ .$$

Since x^2, y^2 freely generate $F\phi$, it is clear that ϕ is a monomorphism. Then we have

$$F_{*F}\ \phi = \langle x, y, t \mid x^2 = t^{-1}xt,\ y^2 = t^{-1}yt \rangle\ .$$

Now put

$$F_n = t^n \langle x, y \rangle t^{-n} \leq F_{*F}F\phi,\ n \in \mathbb{Z}\ ,$$

so that

(i) each F_n, being a conjugate of $F_0 = \langle x, y \rangle \cong F$, is free of rank 2,

and

(ii) $F_n < F_{n+1}$ (*proper* inclusion) for all n as $F_{-1} = \langle x^2, y^2 \rangle$ is properly contained in $\langle x, y \rangle = F_0$.

Finally, put $U = \bigcup_{n \in \mathbb{N}} F_n$, to obtain an example of a locally-free group that is not free (see Example 2 in §3.4). We emphasize that a non-trivial fact has been left unproven here: the validity of (i) and (ii) depends on the crucial fact, referred to above, that the map $G \rightarrow G^*_K\phi$ is an *embedding* (that is, is one-to-one).

The second example is similar to the first, except that we start with $Z = \langle x \mid \rangle$ in place of the free group of rank two. Thus, let

$$G = \langle x, t \mid t^{-1}xt = x^2 \rangle\ ,$$

and consider the normal closure H of $\{x\}$ in G, namely,

$$H = \bigcup_{k \in \mathbb{Z}} \langle t^k x t^{-k} \rangle\ .$$

The obvious Schreier transversal is just $U = \{t^k \mid k \in \mathbb{Z}\}$, whereupon generators and relations for M are given by

$$a_k := t^k x t^{-k},\ t^{k-1}xtx^{-2}t^{-k} = a_{k-1}a_k^{-2},\ k \in \mathbb{Z}\ .$$

Thus

$$H = \langle a_1, a_2, a_3, \ldots \mid a_1 = a_2^2,\ a_2 = a_3^2, \ldots \rangle\ , \tag{11}$$

and there is a homomorphism

$$\left.\begin{array}{l} \theta : H \rightarrow Q \\ a_k \mapsto 2^{-k} \end{array}\right\}$$

into the additive group of rational numbers (Proposition 4.3). Now the relations in (11) show that H is locally cyclic (and in particular, abelian), and any word w in H is equal to a power of a generator. Since $a_k^n \theta = n/2^k \neq 0$ if $n \neq 0$, it follows that θ is one-to-one. Thus, H is isomorphic to the additive group of dyadic rationals $\{n/2^k \mid n \in \mathbb{Z}, \ k \in \mathbb{N}\}$.

As a third and final example, we take the HNN-extension

$$G = <b,t \mid t^{-1}b^2t = b^3> . \tag{12}$$

This is the notorious Baumslag-Solitar group, which is the simplest example of a finitely-generated non-Hopfian group. To see this, consider the factor group

$$K = <b,t \mid t^{-1}b^2t = b^3, \ b = (b^{-1}t^{-1}bt)^2> , \tag{13}$$

obtained by adjoining the relator

$$r = b^{-1}t^{-1}btb^{-1}t^{-1}btb^{-1} .$$

That G is non-Hopfian follows from these two facts:

$$\text{(i) } r \neq e \text{ in } G, \quad \text{(ii) } G \cong K .$$

Now (i) is an immediate consequence of the normal form theorem for HNN-extensions. This is a deep result and beyond our scope. To prove (ii), use Tietze transformations to perform successive adjustment to the pesentation (13) as follows:

$$K = <b,t \mid t^{-1}b^2t = b^3, \ b = (b^{-1}t^{-1}bt)^2>$$

$$X+$$

$$= <b,c,t \mid t^{-1}b^2t = b^3, \ b = c^2, \ c = b^{-1}t^{-1}bt>$$

$$R\pm$$

$$= <b,c,t \mid t^{-1}c^4t = c^6, \ b = c^2, \ c = c^{-2}t^{-1}c^2t>$$

$$X-$$

$$= <c,t \mid t^{-1}c^4t = c^6, \ t^{-1}c^2t = c^3>$$

$$R-$$

$$= <c,t \mid t^{-1}c^2t = c^3> ,$$

which is just the presentation (12) for G.

The Baumslag-Solitar group has a number of remarkable properties, of which we mention one. From (12), it is clear that def $G \geq 1$ (see Definition 7.1). On the other hand, $G_{ab} \cong Z$, so the results of Chapter 6 ensure that def $G \leq 1$. Hence, def $G = 1$. However, in terms of the generators $x = b^4$, $y = t$, G has no one-relator presentation.

8. The Schur multiplicator

The multiplicator $M(G)$ (see Definition 7.2) of a given finite group G can be computed from a finite presentation $G = <X \mid R>$ in the following way. Consider the group

$$G_1 = <X \mid [X,R]>, \quad \text{where } [X,R] = \{[x,r] \mid x \in X, \ r \in R\} \ .$$

The subgroup $H = <R>$ of G_1 is clearly central and thus abelian. Since H is finitely-generated, we have

$$H = A \times T \ ,$$

where A is free abelian and T is finite (see Chapter 6). Then T, the torsion subgroup of G_1, is precisely $M(G)$.

It can be proved that the rank of T is just $|X|$, proving that $M(G)$ can be generated by $|R| - |X|$ elements. The group $\hat{G} = G_1/T$ is called a *covering group* of G. Unlike $M(G)$, $\hat{G}$ is not an invariant of G, essentially because the direct factor A of H is not unique.

The problem of computing $M(G)$ can now be solved:

(i) first use the method of §9.1 to find a presentation of H, and then

(ii) isolate $M(G)$, the torsion subgroup of H, by the methods of Chapter 6.

In the case when G is a finite simple group, the implementation of this requires the use of a high-speed computing machine, and the need to compute $M(G)$ has been the impetus behind the development of much software over the last decade or two. We restrict ourselves to a humbler example, the smallest G for which $M(G)$ is non-trivial.

Let

$$G = Z_2 \times Z_2 = <x,y \mid x^2 = y^2 = (xy)^2 = e> \ ,$$

so that

$$G_1 = <x,y \mid [y,x^2] = [x,y^2] = [x,yxy] = [y,xyx] = e> \ . \tag{14}$$

Taking $U = \{e,x,y,xy\}$, we obtain the following tables.

	x	y				
e	$-$	$-$	$y^{-1}x^{-2}yx^2$	$x^{-1}y^{-2}xy^2$	$x^{-1}y^{-1}x^{-1}y^{-1}xyxy$	$y^{-1}x^{-1}y^{-1}x^{-1}yxyx$
x	x^2	$-$	$xy^{-1}x^{-2}yx$	$y^{-2}xy^2x^{-1}$	$y^{-1}x^{-1}y^{-1}xyxyx^{-1}$	$xy^{-1}x^{-1}y^{-1}x^{-1}yxy$
y	$yxy^{-1}x^{-1}$	y^2	$x^{-2}yx^2y^{-1}$	$yx^{-1}y^{-2}xy$	$yx^{-1}y^{-1}x^{-1}y^{-1}xyx$	$x^{-1}y^{-1}x^{-1}yxyxy^{-1}$
xy	$xyxy^{-1}$	xy^2x^{-1}	$x^{-1}yx^2y^{-1}x^{-1}$	$xyx^{-1}y^{-1}xyx^{-1}$	$xyx^{-1}y^{-1}x^{-1}y^{-1}xy$	$y^{-1}x^{-1}yxyxy^{-1}x^{-1}$

Table 31

$-$	$-$	$\overline{321}30$	$\overline{04}03$	$\overline{0412}3$	$\overline{321}40$
0	$-$	$\overline{412}40$	$\overline{34}$(a)	$\overline{32}014$	$\overline{41}023$
1	3	$\overline{01}2$(b)	$\overline{24}23$(e)	$\overline{23}140$	$\overline{04}132$
2	4	$\overline{02}1$(c)	$\overline{13}14$(d)	$\overline{10}423$	$\overline{32}401$

Table 32

Relators (a), (b) yield that $4 = 3$ and $0 = 12$, whereupon (c), (d), (e) merely assert that $1,2,3$ all commute confirming that H is abelian. The other three relations in the first two columns are now superfluous, and the 8 in the last two columns all reduce to 1^2. It follows that

$$H \cong Z \times Z \times Z_2 \ ,$$

whence $M(G) = Z_2$.

Exercises

1. Find a presentation for A_n in general form. Deduce that $A_5 \cong D(5,3,2)$.

2. By taking $U = \{x_1^n \mid n \in \mathbb{Z}\}$, prove that the subgroup $<a_1,a_2>$ of the group H in (4) is free of rank 2.

3. Let K be the kernel of the homomorphism $v : B_3 \to S_3$. Show that K is the subgroup of index 2 in the group H of (4) generated by a_1, a_2, x_1^2. Prove that $H \cong F_2 \times Z$, where F_2 is free of rank 2.

4.   Show that $(B_n)_{ab} \cong Z$. Find a presentation for B'_n.

5.   Find all solutions of the Diophantine inequality

$$1/l + 1/m + 1/n \geq 1, \text{ with } l \geq m, \geq n \geq 2 .$$

6.   Prove that

$$D(4,4,2)' \cong D(6,3,2)' \cong Z \times Z .$$

7.   Find a presentation for $D(3,3,5)'$.

8.   Consider the group

$$G = \langle x,y \mid x^l = y^m = (xy)^n \rangle .$$

Prove that G_{ab} is infinite if and only if

$$1/n = 1/l + 1/m . \tag{15}$$

Show that every solution of this Diophantine equation has the form

$$(l,m,n) = (hl',hm',bl'm'), \quad h = b(l' + m') ,$$

where b,l',m' are any three positive integers such that $(l',m') = 1$. By exhibiting a suitable homomorphism onto Z, show that $z := x^l$ has infinite order when (15) holds. Deduce that $Z(G)$ is infinite in this case.

9.   Find a 6-generator 1-relator presentation of $D(4,4,4)'$. Reduce this to (6) by Tietze transformations corresponding to surgery on the 2-cell of the following type. Cut across the dodecagon along a segment joining two non-adjacent vertices, labelling the segment (now appearing twice) with a new letter oriented in the direction of the cut. Then paste the two pieces together along a pair of like-oriented edges bearing the same (old) label.

10. Prove that $(Z_{l*}Z_m)'$ is a free group of rank $(l-1)(m-1)$.

11. Let ϕ be an automorphism of a group G. Prove that the HNN-extension $G_{*G}\phi$ is just a semi-direct product $G]Z$ (see §5.4). Show that when ϕ is inner, $G_{*G}\phi \cong G \times Z$.

12. Compute the multiplicator of the symmetric group S_3.

13. Identify the factor group of G_1 in (14) obtained by adjoining the relators corresponding to the generators 2 and 3, namely $xyxy^{-1}$ and y^2. Do the same for the generators 2 and 14. Deduce that both Q_4 and D_4 are covering groups of $Z_2 \times Z_2$.

14. Find a presentation of the subgroup $H = <ab,bc,ca>$ of

$$G = <a,b,c \mid abc = b, \quad bca = c, \quad cab = a> \ .$$

Can you identify G?

15. By finding a presentation for the normal abelian subgroup $H = <x^2,y^2,z^2>$ of

$$J(a,b,c) = <x,y,z \mid x^y = y^{b-2}x^{-1}y^{b+2}, \ y^z = x^{c-2}y^{-1}x^{c+2}, \ z^x = x^{a-2}z^{-1}x^{a+2}>,$$

where a,b,c are non-zero even integers, prove that $J(a,b,c)$ has order $2^8 \mid abc\,(abc - 1)\mid$ (cf. Proposition 7.3).

16. Let $F_3 = <a,b,c \mid>$ and $F_4 = <x,y,z,t \mid>$ be free groups and consider the homomorphisms

$$\alpha : F_3 \ \rightarrow \ <c \mid c^3> \cong Z_3 \ , \qquad \beta : F_4 \ \rightarrow \ <t \mid t^2> \cong Z_2 \ .$$

$$\begin{aligned} a,b &\mapsto e \\ c &\mapsto c \end{aligned}\Bigg\} \qquad\qquad \begin{aligned} x,y,z &\mapsto e \\ t &\mapsto t \end{aligned}\Bigg\}$$

Prove that Ker α, Ker β are freely generated by

$$a, \ b, \ acac^{-1}, \ c^3, \ cbc^{-1}, \ c^2bca^{-1}c^{-3} \ cac^{-2},$$
$$x, \ y, \quad z, \quad t^2, \ txt^{-1}, \quad tyt^{-1}, \quad tzt^{-1},$$

respectively. From the group $\hat{B}_3 := F_3 *_{F_7} F_4$ by identifing these 7 elements in pairs in the given order. Use Tietze transformations to prove that, in terms of the generators $u = c^{-1}t, v = tc^{-1}, w = (ac^{-1}t)^{tc^{-1}}$, we have the presentation

$$\hat{B}_3 = <u,v,w \mid u \approx v \approx w \approx u> \ ,$$

where $\approx$ denotes the braid relation.

CHAPTER 10

PRESENTATIONS OF GROUP EXTENSIONS

1. Basic concepts

Definition 1. An *extension of a group G by a group A* is a group $\tilde{G}$ having a normal subgroup N such that

$$A \cong N , \quad \tilde{G}/N \cong G . \tag{1}$$

This is a very general notion, and its treatment depends very much on one's point of view. (Many authors, for example, would call such a $\tilde{G}$ an extension of A by G). Thus, a group extension arises in each of the following cases:

   (i) given a normal embedding

$$\iota : A \to \tilde{G} ,$$

that is, ι is a monomorphism with Im $\iota \lhd \tilde{G}$, $\tilde{G}$ is an extension of $\tilde{G}/\text{Im } i$ by A;

   (ii) given an epimorphism

$$\nu : \tilde{G} \twoheadrightarrow G ,$$

$\tilde{G}$ is an extension of G by Ker ν.

For our purposes, the best way to picture a group extension is as follows. Name the isomorphisms (1) α, β, respectively, and let ι, ν denote the composites

$$A \xrightarrow{\alpha} N \xrightarrow{\text{inc}} \tilde{G}, \quad \tilde{G} \xrightarrow{\text{nat}} \tilde{G}/N \xrightarrow{\beta} G ,$$

respectively. Then we have a diagram

$$A \xrightarrow{\iota} \tilde{G} \xrightarrow{\nu} G , \tag{2}$$

where the maps satisfy:

$$\text{Ker } \iota = E, \quad \text{Im } \iota = \text{Ker } \nu, \quad \text{Im } \nu = G \ . \tag{3}$$

This prompts the following definition, which will be useful later in another setting.

Definition 2. A sequence

$$A_0 \xrightarrow{\alpha_0} A_1 \xrightarrow{\alpha_1} A_2 \longrightarrow \dots \longrightarrow A_{n-1} \xrightarrow{\alpha_{n-1}} A_n \tag{4}$$

of groups and homomorphisms is called *exact* if

$$\text{Im } \alpha_{i-1} = \text{Ker } \alpha_i, \quad 1 \le i \le n-1 \ .$$

In the case when A_0 and A_n are both trivial and $n = 4$, (4) is called a *short exact sequence* (of groups).

In deference to tradition, we denote the trivial group by 1, rather than E, in this context, and so write a short exact sequence in the form

$$1 \longrightarrow A_1 \xrightarrow{\alpha_1} A_2 \xrightarrow{\alpha_2} A_3 \longrightarrow 1 \ ,$$

where α_0 and α_3, being trivial, have been suppressed. The conditions of exactness at A_1, A_2, A_3 respectively assert that

$$\text{Ker } \alpha_1 = 1, \quad \text{Im } \alpha_1 = \text{Ker } \alpha_2, \quad \text{Im } \alpha_2 = A_3 \ ,$$

which is a restatement of (3). The notions of group extension and short exact sequence of groups are thus *the same*, and we think of an extension as an exact sequence

$$1 \to A \xrightarrow{\iota} \tilde{G} \xrightarrow{\nu} G \to 1 \ .$$

The above ideas form the starting point for (algebraic) homology theory. While we shall not develop this further than we need it (which is not very far), it involves one other simple notion. This we now define (cf. Fig. 1.1) and illustrate by a pretty little lemma.

Definition 3. A *diagram* is a directed graph whose vertices are groups and whose edges are homomorphisms between their endpoints. Such a diagram is called *commutative* if, given any two vertices and any two paths between them, the corresponding composite homomorphisms are equal.

Lemma 1 (The five - lemma). Let

$$
\begin{array}{ccccccccc}
 & \alpha_0 & & \alpha_1 & & \alpha_2 & & \alpha_3 & \\
A_0 & \to & A_1 & \to & A_2 & \to & A_3 & \to & A_4 \\
\downarrow \phi_0 & & \downarrow \phi_1 & & \downarrow \phi_2 & & \downarrow \phi_3 & & \downarrow \phi_4 \\
 & \beta_0 & & \beta_1 & & \beta_2 & & \beta_3 & \\
B_0 & \to & B_1 & \to & B_2 & \to & B_3 & \to & B_4
\end{array}
$$

be a commutative diagram with exact rows. If ϕ_0, ϕ_1, ϕ_3, ϕ_4 are isomorphisms, then so is ϕ_2.

Proof. The proof is by "diagram-chasing", and is easier to see than to write down. We merely give the justification of the successive steps, abbreviating the hypotheses to statements like EA_2 (exactness at A_2) and $C3$ (commutativity of the third square from the left). First, let $a \in \mathrm{Ker}\ \phi_2$, so that $a\phi_2\beta_2 = 1$. Then apply

$$C3,\ \phi_3\ 1-1,\ EA_2,\ C2,\ EB_1,\ \phi_0\ \text{onto},\ C1,\ \phi_1\ 1-1,\ EA_1$$

in turn to deduce that $a = 1$. Hence, ϕ_2 is one-to-one. Next, let $b \in B_2$, so that $b\beta_2 \in B_3$. Then use ϕ_3 onto, EB_3, $C4$, $\phi_4\ 1-1$, EA_3, $C3$, EB_2, ϕ_1 onto, $C2$ in turn to deduce that $b \in \mathrm{Im}\ \phi_2$. Hence, ϕ_2 is onto.

2.   The main theorem

Suppose we are given an extension

$$1 \to A \overset{\iota}{\to} \tilde{G} \overset{v}{\to} G \to 1 \tag{5}$$

and presentations

$$G = \langle X \mid R \rangle, \quad A = \langle Y \mid S \rangle \tag{6}$$

for G and A. Our aim is to put together a presentation for $\tilde{G}$.

First, let

$$\tilde{Y} = \{\tilde{y} = y\iota \mid y \in Y\} ,$$

and let

$$\tilde{S} = \{\tilde{s} \mid s \in S\}$$

be the set of words in $\tilde{Y}$ obtained from S by replacing each y by $\tilde{y}$ wherever it appears.

Next, let

$$\tilde{X} = \{\tilde{x} \mid x \in X\}$$

be members of a transversal for $\text{Im}\, \iota$ in $\tilde{G}$ such that $\tilde{x}\nu = x$ for all $x \in X$. Furthermore, for each $r \in R$, let $\tilde{r}$ be the word in $\tilde{X}$ obtained from r by replacing each x by $\tilde{x}$. Now ν annihilates each $\tilde{r}$, and so for all $r \in R$, $\tilde{r} \in \text{Ker}\, \nu = \text{Im}\, \iota$ and since $\text{Im}\, \iota$ is generated by the set $\tilde{Y}$, each $\tilde{r}$ can be written as a word - say v_r - in the $\tilde{y}$. We put

$$\tilde{R} = \{\tilde{r}v_r^{-1} \mid r \in R\}\ .$$

Finally, since $\text{Im}\, \iota \lhd G$, each conjugate $\tilde{x}^{-1}\tilde{y}\tilde{x}$, $\tilde{x} \in \tilde{X}$, $\tilde{y} \in \tilde{Y}$, belongs to $\text{Im}\, \iota$, and so is a word $- w_{x,y}$ say - in the $\tilde{y}$. Putting

$$\tilde{T} = \{\tilde{x}^{-1}\tilde{y}\tilde{x}w_{x,y}^{-1} \mid x \in X, y \in Y\}\ ,$$

we have the following result.

Proposition 1. With the above notation, the group $\tilde{G}$ has a presentation

$$<\tilde{X}, \tilde{Y} \mid \tilde{R}, \tilde{S}, \tilde{T}>\ . \tag{7}$$

Proof. Letting D be the group presented by (7), it follows from the fact that all the relations in (7) hold in $\tilde{G}$ that there is a homomorphism

$$\left.\begin{array}{c} \theta : D \to \tilde{G} \\ \tilde{x} \mapsto \tilde{x} \\ \tilde{y} \mapsto \tilde{y} \end{array}\right\},$$

by the substitution test. The restriction of θ to the subgroup $<\tilde{Y}>$ of D gives rise to a homomorphism

$$\left.\begin{array}{c} \theta_1 : <\tilde{Y}> \to \text{Im}\, \iota \cong A \\ \tilde{y} \mapsto y \end{array}\right\},$$

and since the defining relations S of A (with each y replaced by $\tilde{y}$) all hold in $<\tilde{Y}> \leq D$, θ_1 must be a bijection. Now the presence of the relations $\tilde{T}$ in (7) means that $<\tilde{Y}>$ is a normal

subgroup of D, and since $<\tilde{Y}>\theta \le \operatorname{Im}\iota$, θ induces a homomorphism

$$\left.\begin{array}{c}\theta_2 : D/<\tilde{Y}> \;\to\; \tilde{G}/\operatorname{Im}\iota \cong G\\[4pt] <\tilde{Y}>\tilde{x} \;\mapsto\; x\end{array}\right\}\quad.$$

Now the relations R defining G all hold (with x replaced by $<\tilde{Y}>\tilde{x}$) in $D/<\tilde{Y}>$, so θ_2 must be a bijection. We thus have a commutative diagram

$$\begin{array}{ccccccccc}
1 & \to & A & \xrightarrow{\iota} & \tilde{G} & \xrightarrow{\nu} & G & \to & 1\\[4pt]
 & & \theta_1\uparrow & & \theta\uparrow & & \theta_2\uparrow & & \\[4pt]
1 & \to & <\tilde{Y}> & \xrightarrow{\text{inc}} & D & \xrightarrow{\text{nat}} & D/<\tilde{Y}> & \to & 1
\end{array}$$

with exact rows. Since θ_1 and θ_2 are isomorphisms, it follows from Lemma 1 that θ also is an isomorphism. This proves the theorem.

Corollary 1. Let $G = <X\,|\,R>$ and $A = <Y\,|\,S>$ be groups, and $\alpha : G \to \operatorname{Aut}A$ a homomorphism such that $y(x\alpha) = w_{x,y}$, a word in $Y^{\pm}$ ($x \in X$, $y \in Y$). Then the semi-direct product $A\,]\,G$ has the presentation

$$A\,]\,G = <X,Y\,|\,R,S,\{x^{-1}yx\,w_{x,y}^{-1}\,|\,x \in X,\, y \in Y\}>\;.$$

Proof. Merely apply the theorem to the group $K = A\,]\,G$ of Definition (5.1), taking $\tilde{y} = (e,y)$, $\tilde{x} = (x,e)$. Then the v_r in (7) are all trivial, and removing all the tildas from (7) gives the result, by (5.6).

Corollary 2. Let $\tilde{G}$ be an extension of G by A. If G and A are finitely presented, then so is $\tilde{G}$.

Proof. Let $\tilde{G}, G, A$ be as in (5), (6) with $|X|$, $|R|$, $|Y|$, $|S|$ all finite. Then (7) is a presentation of $\tilde{G}$ having

$$|X| + |Y|,\qquad |R| + |S| + |X|\,|Y|$$

generators, relators, respectively. Hence, $\tilde{G}$ is finitely presented.

Corollary 3. Let H be a subgroup of finite index in a group G. If H is finitely presented, then so is G.

Proof. This is a consequence of Corollary 2 and Corollary 9.1 (of which it is the converse!) As H is of finite index, it has only finitely many conjugates. Since each of these is of finite index, so is their intersection, call it N. Then N is a subgroup of finite index in H, and so is finitely presented (Corollary 9.2), and G/N is finite, and so is finitely presented (Proposition 4.1). Then G, being an extension of G/N by N, is finitely presented (Corollary 1).

3.   Special cases

As remarked above, the idea of a group extension is a very general one. There are four particularly favourable (and useful) special cases and we now discuss these in turn, all with reference to the extension (5) and its presentation (7).

(S) Semi-direct products

This is the case when $N := \operatorname{Im} \iota$ has a complement, call it C, in $\tilde{G}$, that is, a subgroup $C \le \tilde{G}$ such that

$$\tilde{G} = NC, \quad N \cap C = 1 \tag{8}$$

(cf. formula (5.9)), so that $\tilde{G}$ is a semi-direct product $A]C$. Because of (8), the elements of C form a right transversal for N in G, and it follows that the restriction $v|_C$ is an isomorphism. There is thus a homomorphism σ

$$G \xrightarrow{(v|_C)^{-1}} C \xrightarrow{\text{inc}} \tilde{G}$$

with the property that $v\sigma = 1_G$ (such a σ is called a *splitting* for (5)). Conversely, if (5) is split by $\sigma : G \to \tilde{G}$ (i.e. σ is a homomorphism such that $\sigma v = 1_G$), then $\operatorname{Im} \sigma$ is clearly a complement for N in $\tilde{G}$. It follows that a semi-direct product is nothing other than a split extension.

In the case when (5) is split, by σ say, we can choose the generators $\tilde{X}$ in the proof of Theorem 1 to be $\{x\sigma \mid x \in X\}$. Then for each $r \in R$, we have $\tilde{r} = r\sigma = e$, and in $\tilde{R}$, all the v_r are equal to e. Split extensions of G by A are thus parametrized by the $w_{x,y}, x \in X, y \in Y$, alone. The automorphism α defining the corresponding semi-direct product $A]G$ is then given by

$$\left.\begin{array}{rc} \alpha : G \to & \mathrm{Aut}\, A \\ \\ x \mapsto & \left\{\begin{array}{l} A \to A \\ y \mapsto w_{x,y} \end{array}\right\} \end{array}\right\} . \tag{9}$$

N.B. Starting from presentations (6) and a homomorphism (9), the presentation (7), with every $v_r = e$, always defines an extension of G by A (see Corollary 1). Unfortunately, there appears to be no satisfactory analogue of this in the non-split case.

(A) Extensions with abelian kernel

A is often called the *kernel* of the extension (5). Since $A \lhd \tilde{G}$, there is a homomorphism

$$\gamma : \tilde{G} \to \mathrm{Aut}\, A \tag{10}$$

induced by conjugation. Thus the map $\iota\gamma : A \to \mathrm{Aut}\, A$ is that induced by conjugation within A, and so is trivial if and only if A is abelian. In this case, γ induces (in the sense of Lemma 4.1) a homomorphism $\alpha : G \to \mathrm{Aut}\, A$, as in §5.4. Because of (5.7), this makes A into a G-module in the following sense.

Definition 4. Given a (multiplicatively-written) group G, a (right) G-module is an (additively - written) abelian group A together with an action of G on A (on the right) such that the following axioms hold:

$$\left.\begin{array}{rcl} (a+b)g &=& ag+bg\,, \\ a(gh) &=& (ag)h\,, \\ ae &=& a\,. \end{array}\right\} \tag{11}$$

In this case, there is a satisfactory theory of extensions of G by A (in which conjugation induces the given G-action). Such extensions are classified up to equivalence by the elements of the second cohomology group $H^2(G,A)$. Further discussion of this is beyond our scope, although the notion of G-module will be useful later.

(Z) Central extensions

This is the case when $N = \mathrm{Im}\,\iota$ is contained in the centre of $\tilde{G}$. Thus, not only is A abelian, but the homomorphism γ of (10) is trivial. When this is so, the $w_{x,y}$ appearing in the relators $\tilde{T}$ of (7) are as simple as possible, i.e., $w_{x,y} = x$ for all $x \in X$, $y \in Y$. Central extensions are thus parametrized by the v_r, $r \in R$, alone. In other words, G is determined

by $|R|$ choices from a set with $|A|$ elements, and we have the following lemma.

Lemma 2. The total number of central extensions of a group $G = <X|R>$ by a group A is at most $|A|^{|R|}$.

(D) The direct product

Suppose we are in cases (S) and (Z) at the same time, that is, (5) is a split central extension. Then the v_r in (7) are all equal to e and the $w_{x,y}$ to x, so that (7) reduces to the presentation (4.3) of the direct product $A \times G$. It is not hard to check that this happens if and only if there is a homomorphism $\tau : \tilde{G} \to A$ such that $\iota\tau = 1_A$ (cf. the situation with σ in case (S)).

4.   Finite p– groups

Definition 5. A p-group is a group whose every element has order a power of p, where p is understood to be a prime. Let G be a non-trivial finite p-group. Then it follows from Cauchy's theorem that $|G| = p^n$, some $n \in \mathbb{N}$. It then follows from the class equation that $Z(G) \neq E$, and we have the following elementary result.

Lemma 3. Every non-trivial finite p-group has a central subgroup of order p.

Further properties of finite p-groups will be developed later, but this is all we need for the moment, which is to apply the above ideas to finite p-groups to give rather crude estimates for a) their deficiency, and b) their number.

Proposition 2. Let G be a group of order $p^n, p \in \mathbb{P}, n \in \mathbb{N}$. Then G has a presentation on n generators with $n(n + 1)/2$ defining relations.

Proof. The proof is by induction on n. When $n = 1$,

$$G \cong Z_p = <x|x^p> ,$$

and the result is obvious. Let $n > 1$ and assume the result for groups of order p^{n-1}. By Lemma 3, G has a central subgroup, N say, of order p, and is thus a central extension

$$1 \to N \to G \to G/N \to 1 . \tag{12}$$

The inductive hypothesis guarantees that G/N has a presentation with $n-1$ generators and $n(n-1)/2$ relations. As $N \cong Z_p$, it follows as in Corollary 2 that the presentation (7) for G has $(n-1)+1 = n$ generators and $n(n-1)/2 + 1 + (n-1)1 = n(n+1)/2$ relations, as required.

Proposition 3. The number of groups of order p^n is at most $p^{(n^3-n)/6}$.

Proof. Again we proceed by induction on n, and again the result is obvious when $n=1$. Let $n>1$ and assume the result for groups of order $n-1$. Then G is a central extension (12) with $N \cong Z_p$ an $|G/N| = p^{n-1}$. There are at most $p^{((n-1)^3-(n-1))/6}$ candidates for G/N by the inductive hypothesis, and each has a presentation with $n(n-1)/2$ relations, by Proposition 2. It follows from Lemma 2 that, for a given G/N, there are at most $p^{n(n-1)/2}$ extensions (12). The total number of groups G of order p^n is thus no more than the product of these numbers, that is,

$$p^{((n-1)^3-(n-1))/6} \cdot p^{n(n-1)/2} = p^{(n^3-n)/6},$$

as required.

Exercises

1. Fill in the details of the proof of the five-lemma.

2. Let B_n be the braid group of §9.3 and let β_k be the automorphism of $F_n = \langle a_1,...,a_n | \rangle$ sending a_k to a_{k+1}, a_{k+1} to $a_{k+1}^{-1} a_k a_{k+1}$, and fixing $a_1,...,a_{k-1},a_{k+2},...,a_n$. Prove that the map $\gamma : B_n \to \mathrm{Aut}\, F_n$ sending x_k to β_k, $1 \leq k \leq n-1$, is a homomorphism. Apply Corollary 1 in the case $n=2$ to identify the group (4) of §9.3.

3. Generalize the result of the previous exercise by showing that B_n has a subgroup H ($v^{-1}(\mathrm{Stab}(1))$ in the notation of §9.3) with a presentation (resulting from a Reidemeister-Schreier rewriting with respect to a suitable transversal followed by suitable Tietze transformations) which exhibits H as the semi-direct product $F_{n-1}] B_{n-1}$, where B_{n-1} acts on F_{n-1} via γ (see Exercise 2).

4. Let $\tilde{G} = A] G$, where A and G are residually finite (see Definition 3.5) and A is finitely generated. Use Exercise 2.14 to show that $\tilde{G}$ is residually finite. Deduce that B_n is residually finite (using induction on n, the previous exercise, Exercise 3.15, and Corollary 3.10).

5. Prove that the map inverting the generators of $D(l,m,n)$ is an automorphism. Apply Corollary 1 to deduce that the corresponding split extension $D(l,m,n)]Z_2$ is isomorphic to $\Delta(l,m,n)$.

6. Prove that $\Delta(2,2,n) \cong D(2,2,n) \times Z_2$, for all $n \geq 2$.

7. Identify the finite group $\Delta(2,3,k)$, $k = 3,4,5$.

8. Let G be arbitrary and A *complete*, that is,

$$Z(A) = E, \quad \text{Aut } A = \text{Inn } A \ .$$

Prove that every extension of G by A splits.

9. By adjoining suitable relations, show that the group

$$G = <c_1, c_2, c_3, c_4 \,|\, c_1^2 = c_2^2 = c_3^2 = c_4^2 = e, \ c_1 c_2 c_3 c_4 = c_2 c_4 c_1 c_3 = c_4 c_3 c_2 c_1>$$

has $D_\infty = Z_2 * Z_2$ as a factor group.

10. Show that the group G_1 of the previous exercise has an automorphism of order 5 mapping

$$c \mapsto c_2, c_2 \mapsto c_3, c_3 \mapsto c_4, c_4 \mapsto c_1 c_2 c_3 c_4 \ ,$$

and use Corollary 1 to show that the resulting split extension G_2 of Z_5 by G_1 has a presentation

$$G_2 = <c, d \,|\, c^2 = d^5 = (cd)^5 = (cd^2)^5 = e> \ .$$

11. Prove that the group G_2 of the previous exercise is generated by d and cd^3c.

12. Show that there is a homomorphism from the Fibonacci group $F(2,8)$ onto the group G_2 of the previous exercise, and deduce that $F(2,8)$ is an infinite group.

CHAPTER 11

RELATION MODULES

For the purposes of this section, it will be convenient to regard a presentation of a group G as an extension

$$1 \rightarrow R \xrightarrow{\text{inc}} F \xrightarrow{\phi} G \rightarrow 1 , \tag{1}$$

where F is a free group, ϕ is onto, and $R = \text{Ker } \phi$. The derived group R', being characteristic in R, is normal in F, so that (1) induces an extension

$$1 \rightarrow R/R' \xrightarrow{\text{inc}} F/R' \xrightarrow{\phi} G \rightarrow 1 . \tag{2}$$

(The abuse of notation here is deliberate and will be compounded later: think of ϕ as "going modulo R"). In view of the remarks preceding Definition 10.4 (see §10.3.A), $R_{ab} = R/R'$ thus acquires the structure of a G-module: it is called a *relation module* for G.

Our chief aim here is to construct a short exact sequence of G-modules involving R_{ab} that will be of fundamental importance in the next chapter and in Chapter 15.

1.   *G*-modules

As in Definition 10.4, G-modules will be written additively. In this section, we briefly review various basic (categorical) notions associated with them, and conclude with two examples which will be put together in the next section to yield a third.

A subset B of a G-module of A is a *G-submodule* of A if B is a subgroup of A and is closed under the action of G, that is, $bg \in B$ for all $b \in B$, $g \in G$. Given a submodule B of A, the factor group A/B becomes a G-module if we define

$$(B + a)g = B + ag, \quad a \in A, \ g \in G , \tag{3}$$

called the *factor-module* of A by B.

A *homomorphism* of G-modules is a mapping

$$\theta : A \to B \tag{4}$$

such that

$$(a + b)\theta = a\theta + b\theta, \quad (ag)\theta = (a\theta)g \ ,$$

for all $a,b \in A$, $g \in G$. The *kernel* and *image* of θ are then defined in the usual way:

$$\text{Ker } \theta = \{a \in A \mid a\theta = 0\} \ , \quad \text{Im } \theta = \{a\theta \mid a \in A\} \ ,$$

and these are submodules of A, B respectively.

Given any homomorphism (4) of G-modules, the mapping

$$\left.\begin{array}{lcl} \theta' : & A/\text{Ker } \theta & \to & \text{Im } \theta \\ & \text{Ker } \theta + a & \mapsto & a\theta \end{array}\right\}$$

is an isomorphism of G-modules. Other standard isomorphism theorems hold.

A sequence

$$\dots \to A_{n-1} \xrightarrow{\alpha_{n-1}} A_n \xrightarrow{\alpha_n} A_{n+1} \to \dots$$

of G-modules and homomorphisms is called *exact* if $\text{Im } \alpha_{n-1} = \text{Ker } \alpha_n$ for all n (cf. Definition 10.2). Diagrams of G-modules and their *commutativity* are defined exactly as in Definition 10.3.

Example 1. Any abelian group A can be made into a G-module by defining $ag = a$ for all $a \in A$, $g \in G$. Such G-modules are called *trivial*. The trivial G-module whose underlying group is infinite cyclic is denoted by $\mathbb{Z}$.

Example 2. Given a group G, consider the free abelian group A of rank $|G|$: let $B = \{b_g \mid g \in G\}$ be a basis for A. Let G act regularly on B, that is, $b_g h = b_{gh}$, $\forall g, h \in G$, and extend this to a linear action of G on A. Then, a typical element of A is uniquely of the form

$$a = \sum_{g \in G}{}' n_g b_g \ ,$$

where all the n_g lie in $\mathbb{Z}$ and the ′ denotes that only finitely many of them are non-zero. Then for any $h \in G$,

$$ah = \sum_{g \in G}{}' n_g(b_g h) = \sum_{g \in G}{}' n_g b_{gh} = \sum_{g \in G}{}' n_{gh^{-1}}\, g \ .$$

Note that the multiplication $b_g b_h = b_{gh}$ in B corresponding to that in G extends by distributivity to the whole of A, which thus becomes an associative ring-with-1. We suppress the b's and arrive at the following definitions.

Definition 1. The (integral) *group-ring* $\mathbb{Z}\,G$ of G is the free abelian group with basis G, with multiplication induced by that in G together with the distributive law. The elements of $\mathbb{Z}\,G$ are thus finite integral linear combinations of elements of G, such as

$$\gamma = \sum_{g \in G}{}' n_g g, \quad \delta = \sum_{h \in G}{}' m_h h \ ,$$

and the ring operations are given by

$$\gamma + \delta = \sum_{g \in G}{}' (n_g + m_g)\, g \ ,$$

$$\gamma\delta = \sum_{k \in G}{}' l_k k \ ,$$

where

$$l_k = \sum_{\substack{g,h \in G \\ gh = k}}{}' n_g m_h = \sum_{g \in G}{}' n_g m_{g^{-1}k} \ .$$

Note that a G-module is nothing other than a $\mathbb{Z}\,G$-module in the usual sense of ring theory, that is, the generalization of a vector space in which the scalars lie in an arbitrary ring. The G-notions described above then coincide with the corresponding $\mathbb{Z}\,G$-notions. We explicity mention one of these, as it will be useful later.

Definition 2. A G module P is said to be *free on a subset B* if, for any G-module A and any map $\psi : B \to A$, there is a unique G-homomorphism $\psi' : P \to A$ extending ψ. Then $|B|$ is called the *rank* of P.

The following results are the $\mathbb{Z}\,G$-analogues of Propositions 6.2 and 6.3, and their proofs are left as exercises.

Lemma 1. If P is a free G-module on $B \subseteq P$, then every element of $a \in P$ is uniquely expressible as a finite $\mathbb{Z}\,G$-linear combination of elements of B:

$$a = \sum_{b \in B}' b\gamma_b\,, \quad \gamma_b \in \mathbb{Z}\,G\ .$$

P is thus isomorphic to the direct sum of $|B|$ copies of $\mathbb{Z}\,G$.

Lemma 2. If an arbitrary G-module A is generated by a set of b elements, then A is a G-homomorphic image of $\mathbb{Z}\,G^{\oplus b}$.

2.   The augmentation ideal

Definition 2. The *augmentation mapping* $\varepsilon : \mathbb{Z}\,G \to \mathbb{Z}$ is the homomorphism of abelian groups sending every g in the basis G of $\mathbb{Z}\,G$ to $1 \in \mathbb{Z}$, that is,

$$\left. \begin{array}{ccc} \varepsilon: & \mathbb{Z}\,G & \to & \mathbb{Z} \\ & \displaystyle\sum_{g \in G}' n_g g & \mapsto & \displaystyle\sum_{g \in G}' n_g\ . \end{array} \right\} \tag{5}$$

The kernel of ε is denoted by U and is called the *augmentation ideal* (Magnus ideal, fundamental ideal or difference ideal) of $\mathbb{Z}\,G$.

This terminology is justified by the fact that ε is a homomorphism of rings (Exercise 7), so that U is an ideal of $\mathbb{Z}\,G$. In particular, U is a (right) G-module.

Proposition 1. If X is a generating set for G, then the set $\{x - e \mid x \in X\}$ generates U as a G-module.

Proof. Since $(x - e)\,\varepsilon = x\varepsilon - e\varepsilon = 1 - 1 = 0$, it is clear that the given set is contained in U. On the other hand, if $u = \sum' n_g g$ is in U, then $u\varepsilon = \sum' n_g = 0$, so that

$$u = \sum' n_g g - \sum' n_g e = \sum' n_g\,(g - e)\ ,$$

that is, U is generated as an abelian group by elements of the form $g - e,\ g \in G$ (see Exercise 8). It is therefore sufficient to prove that every $g - e$ is in the G-submodule, call it A, generated by $\{x - e \mid x \in X\}$.

To do this, express g as a word in $X^{\pm}$ and proceed by induction on its length l. The case $l = 0$ is trivial, and for $l = 1$,

$$x - e \in A, \quad x^{-1} - e = -(x - e)x^{-1} \in A .$$

Now let g have length $l > 1$ and assume the result for $l - 1$. Then $g = hx$, where $x \in X^{\pm}$ and $h - e \in A$. But

$$hx - e = (h - e)x + (x - e) ,$$

so that $g - e \in A$, as required.

Now let A be any G-module. Since U is an ideal of $\mathbb{Z}\,G$, the subgroup AU of A generated by the set $\{au \mid a \in A,\ u \in U\}$ is a G-submodule. It is characterized by the fact that A/AU is the biggest G-trivial factor module of A, in the sense of Exercise 9. In the case when A is U itself, we have the following rather striking result.

Proposition 2. The abelian groups U/U^2 and G/G' are isomorphic.

Proof. The composite mapping

$$
\begin{array}{ccccc}
G & \rightarrow & U & \overset{\text{nat}}{\rightarrow} & U/U^2 \\[4pt]
g & \mapsto & g - e &
\end{array}
\Big\}
$$

is a homomorphism since $gh - e$ and $(g - e) + (h - e)$ differ by an element of U^2. Since U/U^2 (additively-written) is abelian, G' lies in the kernel, and we have an induced homomorphism

$$
\alpha:
\begin{array}{ccc}
G/G' & \rightarrow & U/U^2 \\[4pt]
G'g & \mapsto & U^2 + (g - e)
\end{array}
\Big\} \; .
$$

On the other hand, the composite

$$
\begin{array}{ccccc}
U & \rightarrow & G & \overset{\text{nat}}{\rightarrow} & G/G' \\[4pt]
g - e & \mapsto & g &
\end{array}
\Big\}
$$

is a homomorphism sending $(g - e)(h - e) = (gh - e) - (g - e) - (h - e)$ to the coset containing $ghg^{-1}h^{-1} \in G'$, and so it induces a homomorphsim

$$\left.\begin{array}{lll} \beta : & U/U^2 & \to & G/G' \\ & U^2 + (g-e) & \mapsto & G'g \end{array}\right\}$$

which is clearly the inverse of α.

Now suppose we have a free presentation (1) of G, where $F = F(X)$, and assume for the sake of conveniene that X is finite, say $X = \{x_1,...,x_d\}$. Then $X\phi$ generates G, and the mapping

$$\left.\begin{array}{lll} \beta ; & \mathbb{Z}G^{\oplus d} & \to & U \\ & (\gamma_1,...,\gamma_d) & \mapsto & \displaystyle\sum_{i=1}^{d} (x_i\phi - e)\,\gamma_i \end{array}\right\} \tag{6}$$

is an epimorphsim of G-modules, by Proposition 1. Putting $Q = \mathrm{Ker}\,\beta$, we thus have an exact sequence of G-modules

$$0 \to Q \xrightarrow{\mathrm{inc}} \mathbb{Z}G^{\oplus d} \xrightarrow{\beta \circ \mathrm{inc}} \mathbb{Z}G \xrightarrow{\varepsilon} \mathbb{Z} \to 0 . \tag{7}$$

The main aim of this chapter can now be made precise: we will prove that Q is canonically isomorphic to the relation module R_{ab}. The isomorphism will be constructed using the method described in the next section.

3. Derivations

Now let G be any group and A any G-module. Thus, in accordance with Definition 5.1, we can form the semi-direct product $K = A\,]G$ with multiplication

$$(x,a)\,(y,b) = (xy,\ ag + b) .$$

(This is the additive version of (5.8), just as (10.11) is the additive version of (5.7)). Consider a mapping of the form

$$\left.\begin{array}{lll} \theta : & G & \to & K \\ & x & \mapsto & (x,x\partial) \end{array}\right\} ;$$

this is a splitting for the extension

$$0 \to A \to K \to G \to 1$$

if and only if it is a homomorphism of groups, that is, for all $x,y \in G$,

$$(xy,\ (xy)\,\partial) = (xy)\,\theta = x\theta\,y\theta = (x,x\,\partial)(y,y\,\partial) = (xy,x\,\partial y + y\,\partial) ,$$

this is, if and only if ∂ is a derivation, in the following sense.

Definition 3. A map $\partial : G \to A$ from a group G to a G-module A is called a *derivation* if

$$(xy)\,\partial = (x\,\partial)y + y\,\partial, \ \forall\, x,y \in G \ .$$

Derivations $\partial : G \to A$ thus classify complements of A in $A]G$, and we have proved the following result.

Lemma 3. A mapping: $G \to A]G$, $x \mapsto (x, \partial x)$, is a homomorphism if and only if $\partial : G \to A$ is a derivation.

Proposition 1 can now be extended to give a result which will be used later.

Lemma 4. If G is freely generated as a group by X, then the augmentation ideal U of $\mathbb{Z}\,G$ is freely generated as a G-module by the set $X' := \{x - e \,|\, x \in X\}$.

Proof. Let A be any $\mathbb{Z}\,G$-module and $\psi : X' \to A$ any mapping. Then let $\theta' : G \to A]G$ be the homomorphism extending the mapping

$$\theta: \quad \begin{matrix} X & \mapsto & (x, (x-e)\,\psi) \\ x & \to & A]G \end{matrix} \Bigg\} \ .$$

Writing $w\,\theta' = (w, w\,\partial)$, it follows from Lemma 3 that $\partial : G \to A$ is a derivation. We claim that the homomorphism

$$\psi': \quad \begin{matrix} U & \to & A \\ w - e & \mapsto & w\,\partial \end{matrix} \Bigg\}, \ w \in G \setminus \{e\} \ ,$$

is a G-homomorphism extending ψ. Clearly, for $x \in X$,

$$(x, (x-e)\,\psi) = x\theta = x\theta' = (x, x\partial) = (x, (x-e)\,\psi') \ .$$

Moreover, ψ' is additive by definition, and $\forall\, v,w \in G$,

$$((w-e)v)\psi' = ((wv - e) - (v - e))\,\psi'$$

$$= (wv - e)\psi' - (v - e)\psi'$$

$$= (wv)\,\partial - v\,\partial$$

$$= (w\,\partial)v, \ \text{since } \partial \text{ is a derivation} \ ,$$

$$= (w - e)\psi' v .$$

Thus, ψ' commutes with the G-action on a $\mathbb{Z}$-basis and so is a G-homomorphism. Finally, the uniqueness of ψ' follows from Proposition 1.

4.   Free differential calculus

Let $F = F(X)$ be the free group on a set X. For each $x \in X$, there is a mapping

$$\partial/\partial x : F \to \mathbb{Z} F$$

called a *Fox derivative* and computed as follows: given a reduced word w in F, $\partial w/\partial x$ is a sum of terms, one for each occurrence of $x^{\pm 1}$ in w, this term being b when $w = axb$, and $-x^{-1}b$ when $w = ax^{-1}b$. Thus, for example,

$$\partial/\partial x \ (x^{-1}y^{-1}xy) = y - x^{-1}y^{-1}xy , \quad \partial/\partial y \ (x^{-1}y^{-1}xy) = e - y^{-1}xy ,$$

$$\partial/\partial x \ (x^n) = e + x + \ldots + x^{n-1} , \quad \partial/\partial y \ (x^n) = 0 .$$

where $x, y \in X$, $x \neq y$, and $n \in \mathbb{N}$. Specifically, if $w = x_1 \ldots x_n, x_i \in X^{\pm}$, is reduced then

$$\partial w/\partial x = \sum_{i=1}^{n} a_i, \quad \text{where } a_i = \begin{cases} a_{i+1}...a_n , & \text{if } x_i = x , \\ -a_i...a_n , & \text{if } x_i = x^{-1} , \\ 0 , & \text{otherwise} . \end{cases}$$

Remarks 1. This is a right-handed-version of the usual definition, in accordance with our convention of writing homomorphisms, G-actions, etc. on the right.

2. The requirement that w be reduced is not strictly necessary, since the extra terms in $\partial/\partial x$ arising from the insertion of xx^{-1} or $x^{-1}x$ cancel out.

3. The mapping $\partial/\partial x$ extends by linearity to the whole of $\mathbb{Z} F$.

4. (Leibniz' rule). It follows at once from the definition that

$$\frac{\partial}{\partial x}(uv) = \frac{\partial u}{\partial x}v + \frac{\partial v}{\partial x} \tag{8}$$

when there is no cancelling in forming the product of u and v in F. That (8) holds in general follows from Remark 2.

5. $\partial/\partial x$ is characterized by (8) and its values on X:

$$\partial y/\partial x = \delta_{xy}, \quad \forall \, x, y \in X \ ,$$

where δ is the Knonecker δ (merely induct on the length of $w \in F$).

Proposition 3 (The chain rule). If $v_1,...,v_k$ are words in $\{x_1,...,x_d\}^{\pm}$ and w is a word in $\{v_1,...,v_k\}^{\pm}$, then

$$\partial w/\partial x_i = \sum_{j=1}^{k} \partial v_j/\partial x_i \ \partial w/\partial v_j, \quad 1 \le i \le d \ . \tag{9}$$

Proof. We proceed by induction on the length of w as a word in the $v_j^{\pm 1}$, the case of length 0 being trivial. Thus, put $w = v_l^{\varepsilon} w_1$, where $\varepsilon = \pm 1$ and $1 \le l \le k$, and assume the result for w_1. Then, by (8),

$$\partial w/\partial x_i = \partial w_1/\partial x_i + \partial v_l^{\varepsilon}/\partial x_i \ w_1 = \sum_{j=1}^{k} \partial v_j/\partial x_i \ \partial w_1/\partial v_j + \partial v_l^{\varepsilon}/\partial x_i \ w_1 \ .$$

Since $\partial w/\partial v_j = \partial w_1/\partial v_j$ for $j \ne l$, this differs from the desired sum by

$$\partial v_l/\partial x_i \ \partial w_1/\partial v_l - \partial v_l/\partial x_i \ \partial w/\partial v_l + \partial v_l^{\varepsilon}/\partial x_i \ w_1$$

$$= \partial v_l/\partial x_i \ \partial w_1/\partial v_l - \partial v_l/\partial x_i \ (\partial w_1/\partial v_l + \partial v_l^{\varepsilon}/\partial v_l w_1) + \partial v_l^{\varepsilon}/\partial x_i \ w_1 \ ,$$

by (8), and this is zero for either value of ε, since

$$\partial v_l^{-1}/\partial x_i = - (\partial v_l/\partial x_i) v_l^{-1} \ ,$$

by (8) again.

Proposition 4. If $X = \{x_1,...,x_d\}$, then for any $w \in F(X)$,

$$\sum_{i+1}^{d} (x_i - e) \frac{\partial w}{\partial x_i} = w - e \ . \tag{10}$$

Proof. Again proceed by induction on the length of w, the result being trivial when $w = e$. Thus, let $w = x_j^{\varepsilon} w_1$, where $1 \le j \le d$ and $\varepsilon = \pm 1$, and assume the result for w_1. Then by (8),

$$\sum_{i+1}^{d} (x_i - e) \frac{\partial w}{\partial x_i} = \sum_{i+1}^{d} (x_i - e) \left(\frac{\partial w_1}{\partial x_i} + \frac{\partial x_j^{\varepsilon}}{\partial x_i} w_1 \right)$$

$$= w_1 - e + (x_j - e) \frac{\partial x_j^\varepsilon}{\partial x_j} w_1 ,$$

by the inductive hypothesis, and this reduces to $w - e$ in either case, as required.

Note that the formula (10) can be used to *define* the Fox derivatives, since the augmutation ideal of $\mathbb{Z}F$ is *freely* generated (as on F-module) by the $x_i - e$, $1 \leq i \leq d$ (see Lemma 4).

5.   The fundamental isomorphism

We now fix a presentation (1) of G, with $F = F(X)$, $X = \{x_1,...,x_d\}$. The homomorphism ϕ induces a homomorphism of rings

$$\left.\begin{array}{ccc} \phi: & \mathbb{Z}F & \twoheadrightarrow & \mathbb{Z}G \\ & \displaystyle\sum_{w \in F}{}' n_w w & \mapsto & \displaystyle\sum_{w \in F}{}' n_w (w\phi) \end{array}\right\} \tag{11}$$

which in turn induces a homomorphism of abelian groups

$$\left.\begin{array}{ccc} \phi: & \mathbb{Z}F^{\oplus d} & \twoheadrightarrow & \mathbb{Z}G^{\oplus d} \\ & (\gamma_1,...,\gamma_d) & \mapsto & (\gamma_1\phi,...,\gamma_d\phi) \end{array}\right\} \tag{12}$$

The Fox derivatives determine a mapping θ given by

$$\left.\begin{array}{ccc} \partial: & F & \rightarrow & \mathbb{Z}F^{\oplus d} \\ & w & \mapsto & (\partial w/\partial x_1,...,\partial w/\partial x_d) \end{array}\right\} \tag{13}$$

Proposition 5. Consider the composite of the maps (13) and (12):

$$F \xrightarrow{\partial} \mathbb{Z}F^{\oplus d} \xrightarrow{\phi} \mathbb{Z}G^{\oplus d} .$$

Then, for $v \in F$, $v\partial\phi = 0$ if and only if $v \in R'$.

Proof. For the sufficiency, it is enough (by Leibniz' rule) to prove that $(\partial [r,s]/\partial x) \phi = 0$ for all $r,s \in R$ and $x \in X$. Applying (8) three times,

$$\partial [r,s]/\partial x = \partial s/\partial x + \partial r/\partial x \, s + \partial s^{-1}/\partial x \, rs + \partial r^{-1}/\partial x \, s^{-1}rs$$

$$= \partial s/\partial x(1 - s^{-1}rs) + \partial r/\partial x \, (s - r^{-1}s^{-1}rs) ,$$

and this vanishes under ϕ as $r\phi = s\phi = e$.

For the necessity, assume that $(\partial v/\partial x_i) \, \phi = 0,\ 1 \le i \le d$. Since each term on the left is a monomial in $X^{\pm}$ (with coefficient ± 1) and each of these belongs to the $\mathbb{Z}$-basis G of $\mathbb{Z}\,G$, the letters of v are partitioned into pairs with

> a) equal subscript i,   b) opposite sign
>
> c) their contributions to $(\partial v/\partial x_i) \, \phi$ cancelling out.

Thus, each letter $x\ (= x_i)$ in v has a partner x^{-1} such that

$$\begin{array}{llll}
\text{either} & v = axbx^{-1}c & \Rightarrow & \partial v/\partial x = \ldots + (-x^{-1}c + bx^{-1}c) \\
\text{or} & b = ax^{-1}bxc & \Rightarrow & \partial v/\partial x = \ldots + (c - x^{-1}bxc)
\end{array} \Bigg\} ,$$

and condition c) means that $b \in R$ in either case.

Noting that partners cannot be adjacent (v is reduced), let $x^{-\varepsilon}$ be the first letter of v whose partner precedes it, so that if y^{δ} is the letter immediately preceding $x^{-\varepsilon}$, *its* partner must occur later. Thus, there are partners $x^{\pm \varepsilon}$ and $y^{\pm \delta}$ such that v has the reduced form

$$v = a \, x^{\varepsilon} b y^{\delta} x^{-\varepsilon} \, c y^{-\delta} \, d .$$

By what was said in the preceding paragraph, $b y^{\delta}, x^{-\varepsilon} c \in R$, and so

$$v \equiv a x^{\varepsilon} \bullet x^{-\varepsilon} \, c \bullet b y^{\delta} \bullet y^{-\delta} \, d \quad \text{modulo } R'$$

$$= acbd = v', \text{ say } .$$

Thus, $l(v') \le l(v) - 4 < l(v)$ and $v = rv'$ with $r \in R'$. Then, by (8)

$$0 = (\partial v/\partial x_i) \, \phi = (\partial v'/\partial x_i + \partial r/\partial x_i v') \, \phi$$

$$= (\partial v'/\partial x_i) \, \phi + (\partial r/\partial x_i)\phi \, (v'\phi)$$

$$= (\partial v'/\partial x_i) \, \phi ,$$

using the first part of the proof. That $v \in R'$ now follows by induction on $l(v)$.

Proposition 6. Consider the composite of the maps (12) and (6)

$$\mathbb{Z}\,F^{\oplus d} \xrightarrow{\phi} \mathbb{Z}\,G^{\oplus d} \xrightarrow{\beta} U .$$

Then for $\gamma = (\gamma_1, \ldots, \gamma_d) \in \mathbb{Z}\,F^{\oplus d}$, $(\gamma) \, \phi\beta = 0$ if and only if therre is an $r \in R$ with $r\partial\phi = \gamma\phi$.

Proof. The sufficiency follows from Proposition 4: if $r \in R$ satisfies $r\partial\phi = \gamma\phi$, then

$$\gamma\phi\beta = r\partial\phi\beta = (...,\partial r/\partial x_i,...)\,\phi\beta$$

$$= (...,(\partial r/\partial x_i)\,\phi,\,...)\,\beta$$

$$= \sum_{i=1}^{d}(x_i\phi - e)\,(\partial r/\partial x_i)\,\phi$$

$$= (\sum_{i=1}^{d}(x_i - e)\,(\partial r/\partial x_i))\,\phi$$

$$= (r - e)\,\phi = r\phi - e\phi = 0\ .$$

For the necessity, assume that $(\gamma_1,...,\gamma_d)\,\phi\beta = 0$, so that

$$\gamma' := \sum_{i=1}^{d}(x_i - e)\,\gamma_i \tag{14}$$

belongs to the kernel of $\phi : \mathbb{Z}\,F \to \mathbb{Z}\,G$. Thus, the terms in γ' are partitioned into pairs of opposite sign that are equal in G, that is,

$$\gamma' = \sum_{j=1}^{m}(u_j - w_j) = \sum_{j=1}^{m}w_j(r_j - e)\ , \tag{15}$$

where $r_j = w_j^{-1}u_j \in R$, $1 \le i \le m$. Because of Lemma 4, it follows from (14) and (10) that $\gamma_i = \partial\gamma'/\partial x_i$. Thus, from (15),

$$\gamma_i = \sum_{j=1}^{m}(\partial r_j/\partial x_i + \partial w_j/\partial x_i\,(r_j - e))\ ,$$

and so

$$\gamma_i\phi = \sum_{j=1}^{m}(\partial r_j/\partial x_i)\,\phi\ .$$

Putting $r = r_1 r_2 ... r_m \in R$, it follows that

$$(\partial r/\partial x_i)\,\phi = \gamma_i\phi,\quad 1 \le i \le d\ ,$$

i.e., $r\partial\phi = \gamma\phi$ as required.

Theorem 1. With the above notation, the rule

$$\kappa: \begin{array}{ccc} R_{ab} & \to & Q \\ R'r & \mapsto & r\partial\phi \end{array} \Big\}$$

defines an isomorphism of $\mathbb{Z} G$-modules.

Proof. $r\partial\phi \in Q = \mathrm{Ker}\,\beta$ by Proposition 6, and κ is well-defined by Proposition 5. It is a homomorphism by Leibniz' rule (and since $R = \mathrm{Ker}\,\phi$) and $\mathrm{Ker}\,\kappa$ is trivial by Proposition 5. Moreover, κ is onto by Proposition 6 and a G-homomorphism because, for all $g \in G$,

$$((R'r)\,g)\,\kappa = (R'w^{-1}rw)\,\kappa, \quad \text{where } g = w\phi = Rw ,$$

$$= (w^{-1}rw)\,\partial\,\phi$$

$$= (w^{-1}\partial rw + r\partial w + w\partial)\,\phi$$

$$= w^{-1}\partial\phi\,g + r\partial\phi g + w\partial\phi$$

$$= (w^{-1}\partial w + w\partial)\,\phi + r\partial\phi g$$

$$= 0 + (R'r)\,\kappa g ,$$

as required.

Corollary 1. If a group G has a presentation with d generators and r defining relators, then there is an exact sequence of G-modules

$$\mathbb{Z} G^{\oplus r} \xrightarrow{\alpha} \mathbb{Z} G^{\oplus d} \xrightarrow{\beta} R \to 0 . \tag{16}$$

Proof. This will follow from (7) and Lemma 2 provided we can show that $Q \cong R_{ab}$ is generated as a G-module by r elements. But R_{ab} is generated by the F-conjugates of r elements and by definition, the G-action on R_{ab} is that induced by F-conjugation.

Exercises

1. Prove that the action (3) is well defined and satifies the module axioms in Definition 10.4.

2. Let B,C be submodules of a G-module A. Prove that $B/B \cap C$ and $B + C/C$ are isomorphic as G-modules.

3. Let B,C be submodules of a G-module A with $C \subseteq B$. Prove that $(A/C)/(B/C)$ and A/B are isomorphic as G-modules.

4. Let

$$0 \to A \overset{\alpha}{\to} B \overset{\beta}{\to} C \to 0$$

be a short exact sequence of G-modules. Prove that the following three conditions are equivalent:

(i)   there is a G-homomorphism $\sigma : C \to B$ such that $\sigma \beta = 1_C$,

(ii) there is a G-homomorphism $\tau : B \to A$ such that $\alpha \tau = 1_A$,

(iii) B has a G-submodule D such that $D \cap \operatorname{Im} \alpha = 0$ and $D + \operatorname{Im} \alpha = B$.

5. Prove that the addition and multiplication in $\mathbb{Z} G$ given in Definition 1 satisfy the ring axioms.

6. Supply proofs for Lemmas 1 and 2.

7. Prove that the augmentation mapping (5) is both a homomorphism g rings and a homomorphism of G-modules.

8. Prove that U is a free abelian group with basis $\{g - e \mid g \in G, \ g \neq e\}$.

9. Let A be a G-module and B a submodule of A. Prove that A/B is G-trivial if and only if $B \supseteq AU$.

CHAPTER 12

AN ALGORITHM FOR N/N'

As an immediate consequence of the fundamental isomorphism of the previous chapter, we now derive an algorithm for computing the rank and invariant factors of N/N', where N is a normal subgroup of finite index g in a finitely-presented group

$$G = \langle x_1,...,x_d \mid s_1,...,s_r \rangle \ . \tag{1}$$

We also fix the notation

$$\phi: F(X) \twoheadrightarrow G, \quad \psi: G \twoheadrightarrow G/N \ ;$$

these are both natural mappings, and mappings induced by them will also be denoted by ϕ, ψ, respectively. In similar generic fashion, let μ denote "multiplication in G/N". As in the previous chapter, $\partial w/\partial x$ denotes the Fox derivative of $w \in F(X)$ with respect to $x \in X$. Finally, put

$$R = \text{Ker } \phi, \quad S = \text{Ker } \phi\psi \ ,$$

so that we have the following Hasse diagram:

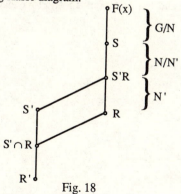

Fig. 18

1. The Jacobian

Definition 1. With the above notation, the *Jacobian* of the presentation (1) is the $r \times d$ matrix J whose (i,j)-entry is $\partial s_i / \partial x_j$.

J is thus a matrix over $\mathbb{Z} F(X)$; its ith row is just $s_i \partial$ in the notation of (11.13). The application of ϕ to each entry yields a matrix $J\phi$ over $\mathbb{Z} G$, and $J\phi\psi$ denotes the corresponding $r \times d$ matrix over $\mathbb{Z}(G/N)$.

Now any matrix over a group-ring can be "blown up" to an integer matrix using the right regular representation of the group. Specifically, let $G/N = \{y_1,...,y_g\}$ and A be a matrix over $\mathbb{Z}(G/N)$. Then any entry γ in A defines a g-tuple of integers by $\gamma = \sum\limits_{k=1}^{g} n_k y_k$. The g-tuple corresponding to γy_l $(1 \le l \le g)$ is rearrangement of this, and we let $\gamma\mu$ denote the $g \times g$ $\mathbb{Z}$-matrix having this as its lth row. Thus, $\gamma\mu$ is a relation matrix for $\mathbb{Z}(G/N)/\gamma\mathbb{Z}(G/N)$ as an abelian group. Finally, let $A\mu$ denote the matrix of integers obtained by applying μ to each entry of A.

Proposition 1. With the above notation, $D = J\phi\psi\mu$ is a relation matrix for the group $N/N' \times \mathbb{Z}^{\times(g-1)}$.

The invariant factors of N/N' can thus be calculated from D by diagonalization (as in Chapter 6), and $rk(N/N')$ is the result of the corresponding computation for ranks, minus $g - 1$.

2. The proof

We begin by recalling some notation and results from the previous chapter. The augmentation ideal U of $\mathbb{Z} G$ is generated as a $\mathbb{Z} G$-module by the $x_j\phi - e$, $1 \le j \le d$, (Proposition 11.1) and

$$\beta : \mathbb{Z} G^{\oplus d} \to U$$

is the corresponding natural $\mathbb{Z} G$-homomorphism (see (11.6)). Now Theorem 11.1 asserts that $Q := \text{Ker } \beta$ is $\mathbb{Z} G$-isomorphic to R_{ab}, and we thus have a short exact sequence

$$0 \to R_{ab} \xrightarrow{\kappa} \mathbb{Z} G^{\oplus d} \xrightarrow{\beta} U \to 0$$

of $\mathbb{Z} G$-modules. Letting V denote the augmentation ideal of $\mathbb{Z}(G/N)$, we have an analogous short exact $\mathbb{Z}(G/N)$-sequence

$$0 \to S_{ab} \xrightarrow{\kappa'} \mathbb{Z}(G/N)^{\oplus d} \xrightarrow{\beta''} V \to 0 .$$

These may be combined to give a commutative diagram with exact rows

$$
\begin{array}{ccccccccc}
0 & \to & R_{ab} & \overset{\kappa}{\to} & \mathbb{Z}\,G^{\oplus d} & \overset{\beta}{\to} & U & \to 0 \\
 & & \downarrow\iota & & \downarrow\psi & & \downarrow\psi & & (2)\\
0 & \to & S_{ab} & \overset{\kappa'}{\to} & \mathbb{Z}\,(G/N)^{\oplus d} & \overset{\beta'}{\to} & V & \to 0
\end{array}
$$

where ι is induced by the inclusion $R \subseteq S$. The commutativity is a consequence of the fact that ψ ("going mod N") commutes with the formation of Fox derivatives (involved in κ and κ').

As noted in the proof of Corollary 11.1, the rows of $J\phi$ generate Im κ over $\mathbb{Z}\,G$, and so the rows of $J\phi\psi$ generate $(\text{Im }\kappa)\,\psi = \text{Im }\kappa\psi = \text{Im }\iota\kappa'$ over $\mathbb{Z}\,(G/N)$. They thus *define* $\mathbb{Z}\,(G/N)^{\oplus d}\,/\,\text{Im }\iota\kappa'$ as a $\mathbb{Z}\,(G/N)$-module. It follows that $D = J\phi\psi\mu$ is a relation matrix for this as an abelian group, call it C. Factoring out Im $i = S'R/S' \cong \text{Im }\iota\kappa'$, this yields a short exact sequence of abelian groups

$$0 \to S_{ab}/(S'R/S') \to C \to V \to 0 \ .$$

Since V is $\mathbb{Z}$-free (of rank $g - 1$) this splits, and so the group defined by D is just the direct sum of $Z^{\times(g-1)}$ and

$$\frac{S/S'}{S'R/S'} \cong S/S'R \cong \frac{S/R}{S'R/R} \cong N/N' \tag{3}$$

(see Fig. 18). This completes the proof.

3.   Examples

We give three examples from earlier chapters, carefully chosen so that manual computation is feasible. In the first, the answer is known already. The second shows that the method may still work when N has infinite index in G. The final example illustrates a useful trick. In all of them we take the favorable special case where $N = G'$.

Example 1. Let G be the free product

$$Z_{l*}\,Z_m = \langle x, y \mid x^l, y^m \rangle \ .$$

Then we know from Exercise 9.10 that G' is a free group of rank $(l-1)(m-1)$.

We first compute the Jacobian J. Let

$$r = x^l, \quad s = y^m$$

in $<x, y \mid >$. Then

$$\partial r/\partial x = 1 + x + \ldots + x^{l-1}, \quad \partial r/\partial y = 0 \ ,$$

$$\partial s/\partial x = 0, \quad \partial s/\partial y = 1 + y + \ldots + y^{m-1} \ .$$

We thus obtain

$$J = \begin{pmatrix} 1 + x + \ldots + x^{l-1} & 0 \\ 0 & 1 + y + \ldots + y^{m-1} \end{pmatrix},$$

and $J\phi\psi$ is the result of regarding this as a matrix over $\mathbb{Z}\, G_{ab}$, which is just the group ring of $Z_l \times Z_m$.

To blow this up to an integer matrix, first multiply $1 + x + \ldots + x^{l-1}$ by the first l powers of x. As it is fixed by each of these, we obtain the $l \times l$ matrix J_l whose every entry is 1. Subsequent multiplication by the first m powers of y yields an $lm \times lm$ matrix with m blocks J_l on the main diagonal and zeroes elsewhere. This is actually the Kronecker product $J_l \otimes I_m$, where I_m is the $m \times m$ identity matrix. The same process applied to $1 + y + \ldots + y^{m-1}$ yields $I_l \otimes J_m$, or equivalently, $J_m \otimes I_l$. Hence,

$$J\phi\psi\mu = \begin{pmatrix} J_l & & & \\ & J_l & & 0 \\ & 0 & J_m & \\ & & & J_m \end{pmatrix},$$

Where there are m J'_ls and l J'_ms.

Since each J_l and J_m disgonalizes to a matrix with a 1 in the top left-hand corner and 0's elsewhere, the first $l + m$ invariant factors of $J\phi\psi\mu$ are 1 and the rest are zero. The rest are $2g - (l + m)$ in number, where $g = |G : G'| = lm$, and it follows from Proposition 1 that $G'/G'' \times Z^{\times(g-1)}$ is free abelian of rank $2lm - (l + m)$. Hence G'/G'' is free abelian of rank

$$2lm - (l + m) - (lm - 1) = (l - 1)\,(m - 1) \ ,$$

as expected.

Example 2. Consider again the Baumslag-Solitar group

$$G = <x,y \mid x^{-1}y^2x = y^3>$$

of § 9.7 (see (9.12)). Putting

$$r = x^{-1}y^{-2}xy^3 \ ,$$

we have

$$\partial r/\partial x = y^3 - x^{-1}y^{-2}xy^3 = y^3 - r \ ,$$

$$\partial r/\partial y = 1 + y + y^2 - y^{-1}xy^3 - y^{-2}xy^3$$

$$= 1 + y + y^2 - yxr - xr \ .$$

Now $r\phi = e$, and so

$$J\phi = (y^3 - e, \ 1 + y + y^2 - yx - x) \ ,$$

and since $y\psi = e$, we obtain

$$J\phi\psi = (0, \ 3e - 2x) \ , \tag{4}$$

where x is regarded as an element of G, G/G', respectively.

In this example, the map β' of (2) has the following effect on a $\mathbb{Z}$-basis:

$$\left. \begin{array}{ccc}
\beta' : & \mathbb{Z} <x \mid> \oplus \mathbb{Z} <x \mid> & \rightarrow & V \\
& (x^k, 0) & \mapsto & (x-1)x^k \\
& (0, x^l) & \mapsto & 0
\end{array} \right\} .$$

Now V is the augmentation ideal of $\mathbb{Z} <x \mid>$, and is thus a free G-module with basis $\{x-1\}$ (Lemma 11.4). The kernel of β' therefore coincides with the second factor, and so $3e - 2x$ is a relation matrix for

$$\frac{\text{Ker } \beta'}{\text{Im } \iota\kappa'} \cong \frac{G'}{G''}$$

(see (3)) as a $\mathbb{Z} <x \mid>$-module. Thus, G'/G'' is defined as an abelian group by

$$(3e - 2x)\mu = \begin{pmatrix} & \cdots & \\ \cdots & 0 \ \ 3 \ -2 \ \ 0 \ \ 0 & \cdots \\ \cdots & 0 \ \ 0 \ \ \ 3 \ -2 \ \ 0 & \cdots \\ & \cdots \ \ \cdots & \end{pmatrix}, \tag{5}$$

where the right-hand side is an infinite circulant matrix.

But the right-hand of (5) is the exponent-sum matrix of the presentation

$$H = <x_k, k \in \mathbb{Z} \,|\, x_k^3 = x_{k+1}^2, \ k \in \mathbb{Z}> \,,$$

so that G'/G'' is isomorphic to H_{ab}. However, the mapping

$$\alpha : \quad \begin{array}{ccc} H & \to & \mathbb{Q}^+ \\ x_k & \mapsto & (2/3)^k \end{array} \Bigg\}$$

clearly extends to a homomorphism, and it can be shown without too much trouble (see Exercises 2 and 3) that

$$\text{Ker } \alpha = H', \ \text{Im } \alpha = \mathbb{Z} \,[1/6] \,, \tag{6}$$

where $\mathbb{Z} \,[1/6]$ is the subgroup $\{k/6^l \,|\, k \in \mathbb{Z}, \ l \in \mathbb{N}\}$ of $\mathbb{Q}^+$. Hence, $G'/G'' \cong \mathbb{Z} \,[1/6]$.

Example 3. Consider again the group

$$G = <x,y \,|\, x^2 yxy^3, \ y^2 xyx^3 >$$

of Example 4.11 and Exercise 8.13, and again take $N = G'$. Then

$$J = \begin{pmatrix} y^3 + yxy^3 + xyxy^3 & 1 + y + y^2 + xy^3 \\ 1 + x + x^2 + yx^3 & x^3 + xyx^3 + yxyx^3 \end{pmatrix},$$

which gives $J\phi$ by regarding x and y as elements of G. Now $G/G' \cong \mathbb{Z}_7$, where

$$x \equiv y, \ x^7 \equiv e \pmod{G'} \,,$$

so that

$$J\phi\psi = \begin{pmatrix} x^3 + x^5 + x^6 & 1 + x + x^2 + x^4 \\ 1 + x + x^2 + x^4 & x^3 + x^5 + x^6 \end{pmatrix}.$$

Before blowing up, we first perform row and column operations on this as a matrix over the ring

$$\mathbb{Z} <x \,|\, x^7> \cong \mathbb{Z}[x]/(x^7 - 1) \,.$$

First, subtract $x^3.$(column 1) from column 2:

$$\begin{pmatrix} x^3 + x^5 + x^6 & 1 + x^4 - x^6 \\ 1 + x + x^2 + x^4 & x^6 - x^4 - 1 \end{pmatrix},$$

then add row 2 to row 1:

$$\begin{pmatrix} 1 + x + x^2 + x^3 + x^4 + x^5 + x^6 & 0 \\ 1 + x + x^2 + x^4 & x^6 - x^4 - 1 \end{pmatrix}.$$

Now blow this up to obtain

$$\begin{pmatrix} J_7 & 0 \\ * & C \end{pmatrix},$$

where J_7 is the 7×7 matrix with 1 in every place, 0 is the 7×7 zero matrix, * is immaterial, and C is the circulant matrix

$$C = \begin{pmatrix} -1 & 0 & 0 & 0 & -1 & 0 & 1 \\ 1 & -1 & 0 & 0 & 0 & -1 & 0 \\ 0 & 1 & -1 & 0 & 0 & 0 & -1 \\ -1 & 0 & 1 & -1 & 0 & 0 & 0 \\ 0 & -1 & 0 & 1 & -1 & 0 & 0 \\ 0 & 0 & -1 & 0 & 1 & -1 & 0 \\ 0 & 0 & 0 & -1 & 0 & 1 & -1 \end{pmatrix}.$$

The invariant factors of J are one 1 and $6 = g - 1$ zeroes. C is thus a relation matrix for G'/G''.

We can now either diagonalize in the usual way, or proceed more efficiently as follows. Observe that C is the exponent-sum matrix of the following cyclically-presented group:

$$H = G_7(x_7 x_5^{-1} x_1^{-1}) .$$

Replacing generators $x_1, ..., x_7$ by their subscripts give the defining relations

$$7 = 15, \; 1 = 26, \; 2 = 37, \; 3 = 41, \; 4 = 52, \; 5 = 63, \; 6 = 74 .$$

Eliminating 7 and 3, by Tietze transformation, we obtain

$$1 = 26, \; 2 = 4115, \; 4 = 52, \; 5 = 641, \; 6 = 154.$$

Similarly removing 2 and 6,

$$1 = 4115154, \ 4 = 54115, \ 5 = 15441.$$

This gives the exponent-sum matrix

$$\begin{pmatrix} 2 & 2 & 2 \\ 2 & 0 & 2 \\ 2 & 2 & 0 \end{pmatrix} \sim \begin{pmatrix} 2 & 0 & 0 \\ 0 & 2 & 0 \\ 0 & 0 & 2 \end{pmatrix},$$

so that $G'/G'' \cong Z_2 \times Z_2 \times Z_2$.

Exercises

1. Consider the special case $N = G$, when ψ is the augmentation mapping $\varepsilon : \mathbb{Z}\,G \to \mathbb{Z}$ and μ is trivial. Prove that, in this case, the matrix $J\phi\psi\mu$ is just the exponent-sum matrix, so that the method reduces to that of Chapter 6 for G/G' in this case.

2. By finding a suitable normal form for the elements of H_{ab} in Example 2, prove that $\text{Ker } \alpha = H'$ (see (6)).

3. Let $a,b \in \mathbb{N}$ be coprime. Prove that

    $$\mathbb{Z}[1/a] \cap \mathbb{Z}[1/b] = \mathbb{Z}, \ \ \mathbb{Z}[1/a] + \mathbb{Z}[1/b] = \mathbb{Z}[1/ab] \ .$$

    Deduce that $\text{Im } \alpha = \mathbb{Z}[1/b]$ (see (6)).

4. Consider the von Dyck group $G = D(l,m,n)$ in the special case when l and m are coprime divisors of n. Prove that the second derived factor of G is the same as that of $Z_l * Z_m$.

5. Let G be the cyclically-presented group

    $$G = \langle x,y \mid x^3 y x y^4, \ y^3 x y x^4 \rangle \ .$$

    Use the method of Example 3 to prove that

    $$G'/G'' \cong F(2,9)_{ab} \cong Z_2 \times Z_{38} \ .$$

6. What can you say about G'/G'' when

    $$g = \langle x,y \mid x^n y x y^{n+1}, \ y^n x y x^{n+1} \rangle \ ,$$

    where $n \in \mathbb{N}$?

CHAPTER 13

FINITE *p*-GROUPS

When considering sets X of generators for a group G, it is important to distinguish between the notions of irredundancy and minimality. Thus, X is *irredundant* if no proper subset of X generates G, and *minimal* if G is generated by no set of smaller cardinality. It is clear that minimality implies irredundancy, and the converse s false in general (Exercise 1). The converse does hold, however, for finite *p*-groups. This is a consequence of the Burnside basis theorem, proved below. The number

$$d(G) := \min \{ |x| \mid \langle x \rangle = G \}$$

thus enjoys enhanced significance as an invariant in the case of finite *p*-groups. We shall be chiefly concerned with this class of groups for the remainder of the book.

1.   Review of elementary properties

By definition, a *p*-group is a group in which every element has order equal to some power of a fixed prime p. It follows from Cauchy's theorem that finite *p*-groups have *p*-power order. It then follows from the class equation that every non-trivial finite *p*-group G has non-trivial centre. This leads to the existence of a *central series* for G, that is, a chain of normal subgroups

$$E = G_0 \leq G_1 \leq \dots \leq G_n = G \tag{1}$$

such that

$$G_k/G_{k-1} \leq Z(G/G_{k-1}),\ 1 \leq k \leq n,$$

or equivalently

$$[G_k, G] \leq G_{k-1}, \quad 1 \leq k \leq n .$$

(For subgroups H,K of a group G, $[H,K]$ denotes the subgroup of G generated by the commutators $\{[h,k] \mid h \in H,\ k \in K\}$). Groups that have a central series are called *nilpotent*.

Lemma 1. Finite p-groups are nilpotent.

Proof. Define a chain of normal subgroups of a finite p-group G inductively as follows:

$$Z_0(G) = E,\quad Z_i(G)/Z_{i-1}(G) = Z(G/Z_{i-1}(G)),\ i \geq 1\ . \tag{2}$$

Every $G/Z_{i-1}(G)$ is a finite p-group, and thus has non-trivial centre unless $Z_{i-1}(G) = G$. As G is finite, the chain $Z_i(G)$, $i = 0, 1, \ldots$, terminates in G after a finite number of steps.

The series just defined is called the *upper central series* of G. The (nilpotency) *class* of G is then the least c such that $Z_c(G) = G$. The *lower central series* of G is defined as follows:

$$\gamma_1(G) = G,\quad \gamma_{i+1}(G) = [\gamma_i(G), G],\ i \geq 1\ . \tag{3}$$

There are, respectively, the fastest ascending and descending central series of a nilpotent group (Exercise 2). It follows that the first trivial term of (3) is $\gamma_{c+1}(G)$.

There is a kind of duality operating here, of which another example is as follows. Minimal normal subgroups of finite p-groups are central of order p (Exercise 3), and the next result is roughly dual to this.

Lemma 2. Every maximal subgroup of a finite p-group G is normal and has index p.

Proof. Let M be a maximal subgroup of G, and (1) a central series. Let l be least such that $M \not\geq G_l$, so that $0 < l \leq n$. Then MG_l is a subgroup of G properly containing M, whence $MG_l = G$ as M is maximal. It follows that $(M/G_{l-1})(G_l/G_{l-1}) = G/G_{l-1}$. Since G_l/G_{l-1} is central in G/G_{l-1}, $M/G_{l-1} \lhd G/G_{l-2}$ and so M is normal in G. Since M is maximal, G/M can have no proper non-trivial subgroup and must therefore have prime order, namely p.

The intersection of all maximal subgroups of any group G is a characteristic subgroup of G called the *Frattini subgroup*, and denoted by $\Phi(G)$. Writing $G^p = <\{g^p \mid g \in G\}>$, we have the following result.

Lemma 3. If G is a finite p-group, then $\Phi(G) = G'G^p$.

Proof. By the preceding lemma, every maximal subgroup M of G contains all pth powers and all commutators. Thus, $G'G^p \subseteq \Phi(G)$. On the other hand, $G/G'G^p$ is an elementary abelian p-group, and so its Frattini subgroup is trivial. $G'G^p$ is thus an intersection of maximal subgroups of G, and so contains $\Phi(G)$.

This motivates the following inductive definition: the series

$$\phi_1(G) = G, \quad \phi_{i+1}(G) = [\phi_i(G), G]\,\phi_i(G)^p, \quad i \geq 1, \tag{4}$$

is called the *lower p-central series* of G. In the case when G is a finite p-group it coincides with the Frattini series, and is the fastest desending central series of G with elementary abelian factors. Such series will be exploited later. Their importance stems from the following fact. Just as abelian groups may be regarded as $\mathbb{Z}$-modules, *elementary* abelian p-groups can be regarded as modules over $GF(p) = \mathbb{Z}/p\mathbb{Z}$ (the naturally induced action is well-defined as every element is annihilated by every multiple of p). But $\mathbb{Z}_p$ is a *field*, and so elementary abelian groups are actually vector spaces, and as such enjoy special properties, such as that invoked in the proof of the next result.

Proposition 1 (Burnside basis theorem). Let X be a subset of a finite p-group G. Then X generates G if and only if the cosets $\{\Phi(G)x \mid x \in X\}$ generate $G/\Phi(G)$. Every irredundant set of generators for G has the same number of elements.

Proof. If $G = \langle X \rangle$, then $G/\Phi(G) = \langle\{\Phi(G)x \mid x \in X\}\rangle$, trivially. For the converse, suppose that X fails to generate G, that is, $H := \langle X \rangle < G$. Then H lies inside some maximal subgroup M of G. But so does $\Phi(G)$, whence $\langle\{\Phi(G)x \mid x \in X\}\rangle \leq M/\Phi(G) < G/\Phi(G)$. It follows that X generates G irredundantly if and only if the corresponding cosets form a basis for $G/\Phi(G)$ (which is a vector space by Lemma 3). Since all bases of a vector space have the same number of elements, the proof is complete.

2. Power-commutator presentations

To obtain a presentation for a given finite p-group G, we can proceed as follows (cf. §10.4). First take a central series

$$G = G_0 > G_1 > \dots G_{n-1} > G_n = E$$

for G with G_i/G_{i-1} cyclic of order p, $1 \le i \le n$, (where $|G| = p^n$). Now choose $a_i \in G_{i-1}/G_i$, $1 \le i \le n$, so that every element $g \in G_i$ ($0 \le i < n$) has a unique normal form

$$g = a_{i+1}^{\alpha_{i+1}} \dots a_n^{\alpha_n}, \text{ where } 0 \le \alpha_j < p, \text{ for } i < j \le n .$$

This is proved by induction, based on the fact that for each i, $1 \le i \le n$, $G_{i-1} = \bigcup_{\alpha=0}^{p-1} a_i^{\alpha} G_i$. Since $a_i^p \in G_i$, $1 \le i \le n$, and $[a_j, a_i] \in G_j$, $1 \le i < j \le n$ (by centrality), it follows that we have equations

$$\text{powers } P : \quad a_j^p = \prod_{k=j+1}^{n} a_k^{\alpha(j,k)}, \quad 1 \le i \le n , \tag{5}$$

$$\text{commutators } C : \quad [a_j, a_k] = \prod_{k=j+1}^{n} a_k^{\alpha(i,j,k)}, \quad 1 \le i < j \le n , \tag{6}$$

where the subscripts increase in both products, and all the α's lie between 0 and $p - 1$. A simple induction based on Proposition 10.1 gives the following result.

Proposition 2. With the above notation,

$$<a_1, \dots, a_n \,|\, P, C> \tag{7}$$

is a presentation of G.

Such a presentation (7) is called a *power-commutator presentation* of G (a PCP, for short).

Example 1. Let G be the metacyclic group

$$<x, y \,|\, y^{p^2} = e = x^p, \ x^{-1}yx = y^{1+p}>$$

of order p^3. Taking

$$G_2 = <y^p>, \quad G_1 = <y>, \quad a_1 = x, \quad a_2 = y, \quad a_3 = y^p ,$$

we obtain the following PCP for G:

$$G = <a_1, a_2, a_3 \,|\, a_1^p = e, \ a_2^p = a_3, \ a_3^p = e,$$

$$[a_2, a_1] = a_3, \ [a_3, a_1] = e, \ [a_3, a_2] = e> .$$

Now suppose, conversely, that we start with a presentation of the form (7) (rather than with a group G), with P and C as in (5) and (6), respectively, where the α's are

chosen arbitrarily in the range $0 \leq \alpha \leq p - 1$. Then (7) defines a group, call it G, in which every element can be expressed in the form

$$\prod_{i=1}^{n} a_i^{\alpha_i} \quad \text{(subscripts increasing), where } 0 \leq \alpha_i \leq p - 1 \text{ for } 1 \leq i \leq n \ . \tag{8}$$

But this form need not be unique (see below), and G may have order less than p^n (but see Exercise 10). Such a PCP is called *consistent* if it indeed defines a group of order p^n.

Example 2. Consider the PCP

$$G = <a_1, a_2, a_3 \mid a_1^p = a_2, \ a_2^p = a_3, \ a_3^p = e \ ,$$

$$[a_2, a_1] = a_3, \ [a_3, a_1] = e, \ [a_3, a_2] = e> \ . \tag{9}$$

Because of the first three relations, G is cyclic and thus abelian. The fourth relation now implies that $a_3 = e$, so that $G \cong Z_{p^2}$.

For the purposes of the next chapter, it will be convenient to have a criterion for deciding when a given PCP is consistent, and we describe one now (without proof). Given a PCP (7), any word w in $\{a_1, \dots, a_n\}^{\pm}$ may be reduced to the form (8) using the relations P and C. This process is called *collection*. However, it may be possible to collect w in many different ways. If the PCP is consistent, all methods of collecting a given w will yield the same normal form, and this will hold for all w in $<a_1, \dots, a_n \mid >$. While the converse of this is true, it provides an unwieldy criterion. We need a smaller set of "test-words", and the following turn out to be sufficient. Each of the words

$$a_k a_j a_i, \quad 1 \leq i < j < k \leq n \ , \tag{10}$$

$$a_j^p a_i, \quad 1 \leq i < j \leq n \ , \tag{11}$$

$$a_j a_i^p, \quad 1 \leq i < j \leq n \ , \tag{12}$$

$$a_i^{p+1}, \quad 1 \leq i \leq n \ , \tag{13}$$

can be collected in two different ways.

Proposition 2 (J.W. Wamsley). A PCP of the form (7) is consistent if and only if the two methods of collecting each of the words (10 - 13) to the form (8) agree in the ranges given.

Example 3. Consider again the group G of Example 2 (see (9)). By Proposition 2, at least one of the 10 words (10 - 13) must collect to two different forms. Take a_1^{p+1}:

$$a_1(a_1^p) = a_1 a_2, \text{ but}$$

$$(a_1^p a_1 = \quad a_2 a_1 = a_1 a_2 a_3 \ ,$$

and these are different.

3.   mod p modules

We now change tack slightly in order to complete the preparations, begun in Chapter 11, for Chapter 15. These consist of "localized forms" of Proposition 11.2 and Corollary 11.1.

First of all, the definition of $\mathbb{Z}\,G$ (Definition 11.1) for a given group G can be generalized to that of the *group-algebra* kG over any commutative ring-with-1 k. In particular, we can take $k = GF(p) = \mathbb{Z}/p\mathbb{Z}$, the field of p elements for some prime p. We fix the prime p and the notation $k = GF(p)$ for the rest of this section. Then kG is a vector space over k of dimension $|G|$.

Alternatively, the subset $p\mathbb{Z}G$ is an ideal of $\mathbb{Z}\,G$, and kG is just the factor ring $\mathbb{Z}G/p\mathbb{Z}G$. In the same spirit, a G-module A determines a kG-module A/pA, since the latter is an elementary abelian p-group, that is, a vector space over k. Moreover any G-homomorphism $\theta : A \to B$ of $\mathbb{Z}\,G$-modules induces a G-homomorphism

$$\left. \begin{array}{rcl} \theta_p : & A/pA & \to & B/pB \\ & a + pA & \mapsto & a\theta + pB \end{array} \right\}$$

of kG-modules. (This clearly commutes with the addition and the G-action, and is well-defined, as $(pA)\,\theta \supseteq pB$).

Proposition 3. If

$$0 \to A \xrightarrow{\theta} B \xrightarrow{\phi} C \to 0$$

is an exact sequence of G-modules with C torison free, then

$$0 \to A/pA \xrightarrow{\theta_p} B/pB \xrightarrow{\phi_p} C/pC \to 0$$

is an exact sequence of kG-modules.

Proof. We check each of the four conditions in turn.

(i) ϕ_p is onto: for any $c \in C$,

$$c + pC = b\phi + pC = (b + pB)\,\phi_p,$$

for some $b \in B$, since ϕ is onto.

(ii) $\operatorname{Im} \theta_p \subseteq \operatorname{Ker} \phi_p$, since $\theta \phi = 0 => \theta_p \phi_p = 0$.

(iii) $\operatorname{Ker} \phi_p \subseteq \operatorname{Im} \theta_p$: for

$$b + pB \in \operatorname{Ker} \phi_p => b\phi + pC = 0$$

$$=> b\phi = pc, \ \text{some } c \in C\ ,$$

$$=> b\phi = p\,(b'\phi), \ \text{some } b \in B, \ \text{since } \phi \ \text{onto}$$

$$=> b - pb' \in \operatorname{Ker} \phi$$

$$=> b - pb' = a\theta, \ \text{since } \operatorname{Ker} \phi \supseteq \operatorname{Im} \theta\ ,$$

$$=> b = a\,\theta + pb' \in \operatorname{Im} \theta_p\ .$$

(iv) θ_p is one-to-one: for

$$a + pA \in \operatorname{Ker} \theta_p => a\theta + pB = 0$$

$$=> a\theta = pb, \ \text{some } b \in B$$

$$=> pb\phi = 0, \ \text{since } \operatorname{Im} \theta \supseteq \operatorname{Ker} \phi\ ,$$

$$=> b\phi = 0, \ \text{since } C \ \text{torsion free}\ ,$$

$$=> a\theta = pb = pa'\theta, \ \text{since } \operatorname{Ker}\phi \subseteq \operatorname{Im}\theta\ ,$$

$$=> a - pa' \in \operatorname{Ker} \theta$$

$$=> a = pa', \ \text{since } \theta \ \text{one-to-one}\ ,$$

$$=> a + pA = 0 \ \text{in } A/pA\ .$$

Corollary 1. There is a short exact sequence of kG-modules

$$0 \to U/pU \xrightarrow{\ \varepsilon_p\ } kG \to k \to 0\ . \tag{14}$$

Proof. Apply the proposition to

$$0 \to U \to \mathbb{Z}G \overset{\varepsilon}{\to} \mathbb{Z} \to 0$$

(see (11.5)).

Corollary 2. If G has a presentation with d generators and r defining relators, there is an exact sequence

$$kG^{\oplus r} \overset{\alpha_p}{\to} kG^{\oplus d} \overset{\beta_p}{\to} U/pU \to 0 \ . \tag{15}$$

Proof. Break up the sequence (11.16) into two short exact sequences

$$0 \to \operatorname{Ker} \alpha \to \mathbb{Z}G^{\oplus r} \to \operatorname{Im} \alpha \to 0 \ ,$$

$$0 \to \operatorname{Ker} \beta \to \mathbb{Z}G^{\oplus d} \to U/pU \to 0 \ ,$$

apply the proposition to both, then splice them together again.

The augmentation mapping ε_p of (14) merely adds together the coefficients (mod p) of each $\gamma \in kG$, and its kernel U/pU (the augmentation ideal of kG) has the k-basis $\{g - e \mid g \in G \setminus \{e\}\}$.

Proposition 4. $\qquad (U/pU)/(U/pU)^2 \cong G/G'G^p \ . \tag{16}$

Proof. Using Proposition 11.2,

$$(U/pU)/(U/pU)^2 = (U/pU)/(U^2 + pU/pU)$$

$$\cong U/U^2 + pU$$

$$\cong (U/U^2)/(pU + U^2/U^2)$$

$$= (U/U^2)/p(U/U^2)$$

$$\cong (G/G')/(G/G')^p$$

$$= (G/G')/(G^p G'/G')$$

$$\cong G/G'G^p \ .$$

Our localization process is now complete. From now on, we work solely with kG-modules and, for the sake of convenience, we let U denote the augmentation ideal of kG. For any kG-module A, AU is then a kG-submodule of A and A/AU admits a kG-module structure (which is in fact G-trivial; see Exercise 12). Moreover, if $\theta : A \to B$ is any kG-homomorphism, then $(AU)\theta \subseteq BU$, and so θ induces a homomorphism

$$\theta': \quad \left.\begin{array}{ccc} A/AU & \to & B/BU \\ a + AU & \mapsto & a\theta + BU \end{array}\right\} .$$

We can now summarize everything we need for Chapter 15 into one proposition.

Proposition 5. Let G be a finite p-group presented with $d = d(G)$ generators and r relations. Then there is an exact sequence of kG-modules

$$A \overset{\alpha}{\to} B \overset{\beta}{\to} U \to 0 , \tag{17}$$

where A and B are free of ranks r and d, respectively, and Im $\alpha \subseteq BU$.

Proof. This is just the sequence (15) in a different notation, and so we merely have to prove the last assertion. Consider the induced sequence

$$A/AU \overset{\alpha'}{\to} B/BU \overset{\beta'}{\to} U/U^2 \to 0 .$$

Here,

$$B/BU = kG^{\oplus d} / U^{\oplus d} \cong k^{\oplus d} \cong U/U^2 ,$$

by Propositions 1 and 4 (since $d = d(G)$). Moreover, β' is onto since β is, and is thus an isomorphism by the pigeon-hole principle. Finally, $\alpha'\beta' = 0$ since $\alpha\beta = 0$, whence

$$A\alpha + BU/BU = \text{Im } \alpha' \subseteq \text{Ker } \beta' = 0 ,$$

so that $A\alpha \subseteq BU$, as required.

Exercises

1. Write down a set of generation for the symmetric group S_4 that is irredundant but not minimal.

2. Prove that any central series (1) of a nilpotent group G satisfies

$$\gamma_{n-i+1}(G) \leq G_i \leq Z_i(G), \quad 0 \leq i \leq n \ .$$

3. Let N be a non-trivial normal subgroup of a nilpotent group G. Show that $N \cap Z(G) \neq E$. Deduce that every minimal normal subgroup of a finite p-group is central and has order p.

4. Let G be a finite p-group and $H < G$. Use a counting argument on double cosets to prove that $H < N_G(H)$, and thus give an alternative proof of Lemma 2.

5. Prove that the Frattini subgroup of an arbitrary group G consists precisely of the "non-generators" of G, that is, $g \in \Phi(G)$ if and only if g has the following property: whenever a subset s generates G, so does $s \setminus \{g\}$.

6. Let G be a finite p-group and let

$$1 \to Z_p \to \tilde{G} \to G \to 1$$

be any extension of G by Z_p. Prove that either

$$\tilde{G} \cong G \times Z_p \quad \text{or} \quad d(\tilde{G}) = d(G) \ .$$

7. Let G be a group of order p^n. Prove that G has a normal series

$$E = G_0 < G_1 < ... < G_n = G \ .$$

The factor G_i/G_{i-1} is said to be complemented if it has a complement in G/G_{i-1}. Show that the number of complemented factors in any such series is equal to $d(G)$.

8. Let $G = <X \mid R>$ be a finite p-group with $|X| = d(G)$. prove that the exponent-sum of every generator in every defining relator is a multiple of p.

9. Give a detailed proof of Proposition 2.

10. Prove (by induction on n) that a *PCP* of the form (7) always defines a p-group.

11. Collect each of the other nine words (10 - 13) in Example 3 in two different ways and check which of them give the same normal form.

12. Prove that A/AU is the largest G-trivial factor-module of A in the following sense: given a kG-submodule B of A,

$$A/B \text{ is } G\text{-trivial} \Leftrightarrow AU \subseteq B \ .$$

CHAPTER 14

THE NILPOTENT QUOTIENT ALGORITHM

The aim of the nilpotent quotient algorithm is to derive, from a given finite presentation $G = <X \mid R = e>$, consistent power-commutator presentations for prime-power factor groups of G. When G is finite, these may be combined (see Exercise 3) to give all the nilpotent factor groups of G, in particular the biggest one $G / \bigcap_{j \in \mathbb{N}} \gamma_k(G)$. In practice, we always work with a fixed prime p, where only those dividing the order of G_{ab} are relevant. The algorithm is an operator

$$NQA : CPCP\,(G/\phi_k(G)) \to CPCP\,(G/\phi_{k+1}(G)) \ ,$$

which thus successively yields the factors by the terms of the lower p-central series (see (13.4)) in $CPCP$ form. The method was devised by I.D. Macdonald in the arly '70s, when the use of high-speed computing machines in Group Theory was fairly well-established, and is less suitable than the methods described in Chapters 8 and 12 for manual computation.

1. The algorithm

The input at the kth stage consists of the original presentation

$$G = <x_1,...,x_m \mid R = e>$$

together with a consistent PCP (of the form (13.5-7)) for $G/\phi_k(G)$, with the property (guaranteed inductively) that for each i, $1 \le i \le n$, there is a relation in (13.5 or 6) containing a_i exactly once, at the right-hand end of the right-hand side. This relation is called the *definition* of a_i and is suitably labelled D. The combined input is depicted by the scheme in Fig. 19.

There are four types of relation involved, namely, types C,P,R,X, in number $\binom{n}{2}$, n, $\mid R \mid$, m, respectively. The presentation

$$<a_1,...,a_n \,|\, C,P>$$

is a *CPCP* for $G/\phi_k(G)$. Among relations of types C,P,X, there are exactly n D's, of which the first d, where $|G : G'G^p| = p^d$, are of type X and we usually relabel $x_i = a_i$, $1 \le i \le d$.

$$a_1,...,a_n \qquad\qquad\qquad\qquad\qquad x_1 ... x_m \,|\, R = e$$

$$x_i = \prod_j a_j^{\zeta(i,j)}, \quad 1 \le i \le m$$

$$\begin{cases} 1 \le i < j \le n, & [a_j,a_i] = \\ 1 \le j \le n, & a_j^p = \end{cases} \left.\begin{matrix} \\ \end{matrix}\right\} \prod_{j+1}^{n}, \text{ as in (13.6,5)}$$

$$\underbrace{\hphantom{xxxxxxxxxxxxxxxxxxxxxxx}}_{C,P} \qquad\qquad \underbrace{\hphantom{xxxxxxxxxxxxx}}_{X}$$

$$T : t\text{'s central of order } p$$

Fig. 19

At the start of Stage 1 there are no a's ($n = 0$), P and C are empty, and the relations of type X are just $x_i = e$, $1 \le i \le m$, which presents the trivial group $G/\phi_1(G)$. At the start of stage 2, $n = d$, and the d D's are taken as the first d relations of type X.

The algorithm now proceeds in four steps:

1. Stick a temporary new generator on the right-hand end of every relation of type C,P,X which is not already a definition D; call these

$$\{t_j \,|\, 1 \le j \le \binom{n}{2} + m\},$$

and carry temporary relations T which assert that the t's all have order p and commute with each other and the a's:

$$t_j^p = e, \quad 1 \le j \le \binom{n}{2} + m,$$

$$[t_j,t_i] = e, \quad 1 \le i < j \le \binom{n}{2} + m,$$

$$[t_j,a_i] = e, \quad 1 \le j \le \binom{n}{2} + m, \quad 1 \le i \le n.$$

2. Ensure that the presentation is consistent: this involves collecting each of the words (13.10-13) in two ways, and results in a set E of equations in the t's.

3. Enforce the relations $R = e$: this involves substitution of the relations of type X in R and then collecting the resulting words in $\{a_1,...,a_n\}^{\pm}$ into normal form (increasing subscripts). This results in another set E' of equations in the t's.

4.   Solve the homogeneous system $E \cup E'$ of linear equations over $GF(p)$, and thus express all t's in terms of some subset that is linearly indpendent modulo $E \cup E'$. Relabel this subset $\{a_{n+1}, ..., a_{n'}\}$ and subsitute for all t's throughout. Adjust the remaining T-relations to PCP form and adjoin them to the original P and C relations. Finally, attach the label D to the $n' - n$ relations where the t's corresponding to $\{a_{n+1}, ..., a_{n'}\}$ were originally placed.

The result is a $CPCP$ for $G/\phi_{k+1}(G)$, and forms the input for the $k + 1$st stage. The process terminates when all the t's vanish in Step 4 ($n = n'$). If this happens first at the kth stage, then $\phi_k(G) = \phi_{k+1}(G)$, and we have reached the p-terminal: $G/\phi_k(G)$ is the maximal p-quotient group of G.

2.   An example

Let G be the undistingnished group of order 16 (see Exercise 5.14)

$$G = \langle x_1, x_2, x_3 \mid x_1^2 = x_2^2 = x_3^2 = e, \ x_2 x_3 x_1 = x_1 x_2 x_3 = x_3 x_1 x_2 \rangle$$

and (of course) $p = 2$. We perform stages, and steps within stages, in turn.

Stage 1. The input is just the given presentation together with the equations $x_i = e$, $1 \le i \le 3$.

1.   Step 1 results in the scheme shown in Fig. 20.

$$x_1, x_2, x_3 \mid x_1^2 = x_2^2 = x_3^2 = e$$

$$x_2 x_3 x_1 = x_1 x_2 x_3 = x_3 x_1 x_2$$

$$x_i = t_i, \ 1 \le i \le 3 \quad t_i^2 = e, \ 1 \le i \le 3$$

$$[t_j, t_i] = e, \quad 1 \le i < j \le 3$$

Fig. 20

2.   This is trivial, as there are no a's.

3.   The R-relations (in terms of the t's) are consequences of the T-relations, so again there is no informatioon.

4.   Since E and E' are empty, the t's are linearly independent. Relabel $t_i = a_i$, $1 \leq i \leq 3$, eliminate all t's, convert T-relations into C- and P-relations and insert D's. This results in the input for stage 2, as shown in Fig. 21.

Stage 2.

a_1, a_2, a_3 $\qquad$ $x_1, x_2, x_3 \, | \, R = e$, $\quad$ as in Fig. 20

$a_1^2 = e$ $\qquad\qquad$ $x_1 = a_1 \;\; D$

$a_2^2 = e$ $\qquad\qquad$ $x_2 = a_2 \;\; D$

$a_3^2 = e$ $\qquad\qquad$ $x_3 = a_3 \;\; D$

$[a_2, a_1] = e$

$[a_3, a_1] = e$

$[a_3, a_2] = e$

Fig. 21

1.   Insert t's into Fig. 21 to give Fig. 22.

a_1, a_2, a_3 $\qquad\qquad$ $x_1, x_2, x_3 \;\; | \;\; x_1^2 = x_2^2 = x_3^2 = e$

$\qquad\qquad\qquad\qquad\qquad$ $x_2 x_3 x_1 = x_1 x_2 x_3 = x_3 x_1 x_2$

$a_1^2 = t_1$

$a_2^2 = t_2$ $\qquad\qquad$ $x_1 = a_1 \;\; D$

$a_3^2 = t_3$ $\qquad\qquad$ $x_2 = a_2 \;\; D$

$[a_2, a_1] = t_4$ $\qquad\quad$ $x_3 = a_3 \;\; D$

$[a_3, a_1] = t_5$ $\qquad\qquad\qquad$ $[t_j, t_i] = e$, $\;\; 1 \leq i < j \leq 6$

$[a_3, a_2] = t_6$ $\qquad\qquad\qquad$ $[t_j, a_i] = e$, $\;\; 1 \leq j \leq 6$, $1 \leq i \leq 3$

$\qquad\qquad\qquad\qquad\qquad\;\;$ $t_j^2 = e$, $\;\; 1 \leq j \leq 6$

Fig. 22

2.   We collect the words (13.10-13) on the left and right as indicated below

$$a_3 a_2 a_1$$

$$a_2 a_3 t_6 a_1 \qquad\qquad a_3 a_1 a_2 t_4$$

$$a_2 a_1 a_3 t_5 t_6 \qquad\qquad a_1 a_3 t_5 a_2 t_4$$

$$a_1 a_2 a_3 t_4 t_5 t_6 \qquad\qquad a_1 a_3 a_2 t_4 t_5$$

$$a_1 a_2 a_3 t_4 t_5 t_6$$

$a_2^2 a_1$		$a_3^2 a_1$		$a_3^2 a_2$	
$t_2 a_1$	$a_2 a_1 a_2 t_4$	$t_3 a_1$	$a_3 a_1 a_3 t_5$	$t_3 a_2$	$a_3 a_2 a_3 t_6$
$a_1 t_2$	$a_1 a_2^2 t_4^2$	$a_1 t_3$	$a_1 a_3^2 t_5^2$	$a_2 t_3$	$a_2 a_3^2 t_6$
	$a_1 t_2$		$a_1 t_3$		$a_2 t_3$

$a_2 a_1^2$		$a_3 a_1^2$		$a_3 a_2^2$	
$a_1 a_2 t_4 a_1$	$a_2 t_1$	$a_1 a_3 t_5 a_1$	$a_3 t_1$	$a_2 a_3 t_6 a_2$	$a_3 t_2$
$a_1^2 a_2 t_4^2$		$a_1^2 a_3 t_5^2$		$a_2^2 a_3 t_6^2$	
$a_2 t_1$		$a_3 t_1$		$a_3 t_2$	

a_1^3		a_2^3		a_3^3	
$t_1 a_1$	$a_1 t_1$	$t_2 a_2$	$a_2 t_2$	$t_3 a_3$	$a_3 t_3$
$a_1 t_1$		$a_2 t_2$		$a_3 t_3$	

The resulting information is trivial.

3. The first three R-relations immediately yield $t_1 = t_2 = t_3 = e$. Collecting the other two R-relations, we obtain

$$a_1 a_2 a_3 = x_1 x_2 x_3 = x_2 x_3 x_1 = a_2 a_3 a_1$$

$$= a_2 a_1 a_3 t_5 = a_1 a_2 t_4 a_3 t_5 = a_1 a_2 a_3 t_4 t_5 \,,$$

$$a_1 a_2 a_3 = a_3 a_1 a_2 = a_1 a_3 t_5 a_2 = a_1 a_3 a_2 t_5$$

$$= a_1 a_2 a_3 t_6 t_5 \,,$$

which together yield $t_4 = t_5 = t_6$.

4. From the information just obtained, we put

$$t_1 = t_2 = t_3 = e, \quad t_4 = t_5 = t_6 = a_4 \,.$$

By adjusting Fig. 22 accordingly, we obtain the input for Stage 3 shown in Fig. 23.

Stage 3

a_1, a_2, a_3, a_4	$x_1, x_2, x_3 \mid R = e$ as in Fig. 22	$[a_2,a_1] = a_4 \ D$
$a_1^2 = e$		$[a_3,a_1] = a_4$
$a_2^2 = e$	$x_1 = a_1 \ D$	$[a_3,a_2] = a_4$
$a_3^2 = e$	$x_2 = a_2 \ D$	$[a_4,a_1] = e$
$a_4^2 = e$	$x_3 = a_3 \ D$	$[a_4,a_2] = e$
		$[a_4,a_3] = e$

Fig. 23

1. The result of applying step 1 is to insert T-relations in Fig. 23 and to replace the ten P- and C-relations by the equations

$$a_1^2 = t_1,\ a_2^2 = t_2,\ a_3^2 = t_3,\ a_4^2 = t_4,\ [a_2,a_1] = a_4\ ,$$

$$[a_3,a_1] = a_4t_5,\ [a_3,a_2] = a_4t_6,\ [a_4,a_1]) = t_7,\ [a_4,a_2] = t_8,\ [a_4,a_3] = t_9\ .$$

2. We save space by indicating only the words which yield significant information: collecting

$$a_3a_2a_1,\ a_3^2a_2,\ a_3a_1^2,\ a_3a_2^2$$

yields, respectively,

$$t_7t_8t_9 = e,\ t_4t_9 = e,\ t_4t_7 = e,\ t_4t_8 = e\ . \tag{1}$$

For example,

$$\begin{aligned} a_3(a_2a_1) &= a_3a_1a_2a_4 = a_1a_3a_4t_5a_2a_4 \\ &= a_1a_3a_2a_4t_8a_4t_5 \\ &= a_1a_2a_3a_4t_6a_4t_8a_4t_5 \\ &= a_1a_2a_3a_4t_4t_5t_6t_8\ , \end{aligned}$$

whereas

$$\begin{aligned} (a_3a_2)\,a_1 &= a_2a_3a_4t_6a_1 \\ &= a_2a_3a_1a_4t_7t_6 \\ &= a_2a_1a_3a_4t_5a_4t_7t_6 \end{aligned}$$

$$= a_1 a_2 a_4 a_3 a_4^2 t_5 t_7 t_6$$

$$= a_1 a_2 a_3 a_4 t_9 a_4^2 t_5 t_7 t_6$$

$$= a_1 a_2 a_3 a_4 t_4 t_5 t_6 t_7 t_9 \ ,$$

which yields $t_8 = t_7 t_9$, or $t_7 t_8 t_9$, as claimed

3. The first three R-relations yield

$$t_1 = t_2 = t_3 = e \ , \tag{2}$$

as before. Collecting the other two,

$$a_1 a_2 a_3 = a_2 a_3 a_1 = a_2 a_1 a_3 a_4 t_5 = a_1 a_2 a_4 a_3 a_4 t_5$$

$$= a_1 a_2 a_3 a_4 t_9 a_4 t_5 = a_1 a_2 a_3 t_4 t_5 t_9 \ ,$$

and

$$a_1 a_2 a_3 = a_3 a_1 a_2 = a_1 a_3 a_4 t_5 a_2 = a_1 a_3 a_2 a_4 t_8 t_5$$

$$= a_1 a_2 a_3 a_4 t_6 a_4 t_8 t_5 = a_1 a_2 a_3 t_4 t_5 t_6 t_8 \ ,$$

yielding

$$t_4 t_5 t_9 = e = t_4 t_5 t_6 t_8 \tag{3}$$

4. Using (2), (1), (3) in turn, we see that all the t_i are equal to e. The process is thus finished, and the maximal 2-factor group of G has the PCP

$$G / \phi_3(G) = \ <a_1, \ a_2, \ a_3, \ a_4 \ | \ a_1^2 = a_2^2 = a_3^2 = a_4^2 = e,$$

$$[a_4, a_1] = [a_4, a_2] = [a_4, a_3] = e, \ [a_2, a_1] = [a_3, a_1] = [a_3, a_2] = a_4> \ .$$

3.   An improvement

The above example reinforces the comment made earlier on the unsuitability of the method for hand calculations. The amount of work involved at each stage depends upon the number n of a's, and this increases from stage to stage, as does the complexity of the collecting process. Most of the work is in Step 2, specifically in collecting the words $a_k a_j a_i$ of (13.10), which are $O(n^3)$ in number. A substantial saving, especially for later stages, is achieved by the following result (stated without proof) which reduces $O(n^3)$ to $O(dn^2)$.

Proposition 1. (M.R. Vaughan-Lee). In carrying out Step 2 at any stage, it is sufficient to collect the words (13.10,13) for $1 \leq i \leq d$.

Exercises

1. Prove that a direct product of finitely many nilpotent groups is nilpotent.

2. Prove that, in a nilpotent group, every subgroup is properly contained in its normalizer.

3. Using the previous two exercises and Sylow's theorem, prove that a finite group is nilpotent if and only if it is the direct product of its Sylow subgroups.

4. Prove that the exponent of each factor in the lower central series of any group divides that of its predecessor.

5. Show that the only primes p for which a given G has a non-trivial p-quotient are those dividing $|G_{ab}|$.

6. Use the Burnside basis theorem to prove that if a prime p divides $|G_{ab}|$ to the first power only, then the lower p-central series of G terminates after one step in a subgroup of index p.

7. Using the result of Exercise 12.4 and the previous exercise as a starting point, apply the nilpotent quotient algorithm to the Fibonacci group

$$F(2,9) = G_9(x_1 x_2 x_3^{-1})$$

    to show that the maximal nilpotent quotient of this group is isomorphic to $Z_{19} \times Q_4$.

8. Investigate the 11-quotient structure of the group

$$F(2,10) = G_{10}(x_1 x_2 x_3^{-1}) \; .$$

CHAPTER 15

THE GOLOD-SHAFAREVICH THEOREM

Theorem 1. Let $G = \langle X \mid R \rangle$ be a finite p-group with $|X| = d = d(G)$ minimal and $|R| = r$. Then $r > d^2/4$.

1. The proof

The proof given here is due to P. Roquette and is based solely on the exact sequence (13.17),

$$A \stackrel{\alpha}{\to} B \stackrel{\beta}{\to} U \to 0 \tag{1}$$

of kG-modules, where $k = GF(p)$, U is the augmentation ideal of kG, A and B are kG-free of ranks r and d, respectively, and

$$\operatorname{Im} \alpha \subseteq BU \ . \tag{2}$$

We now proceed in three steps.

(i)   For $l = 0,1,2,...,$ define

$$A_l = \alpha^{-1}(BU^l) \ .$$

Since $A_l \alpha \subseteq BU^l$ and $(BU^l)\beta \subseteq U^{l+1}$, we have a sequence

$$0 \to A_l/A_{l+1} \stackrel{\alpha_l}{\to} BU^l/BU^{l+1} \stackrel{\beta_l}{\to} U^{l+1}/U^{l+2} \to 0 \tag{3}$$

for each $l \geq 0$, where α_l and β_l are induced by α and β, respectively. It is clear that α_l is one-to-one and that β_l is onto. Furthermore,

$$\operatorname{Ker} \beta_l = \frac{BU^{l+1} + \operatorname{Ker} \beta \cap BU^l}{BU^{l+1}} \ ,$$

and

$$\text{Im } \alpha_l = \frac{BU^{l+1} + \text{Im } \alpha \cap BU^l}{BU^{l+1}} \, ,$$

so the exactness of (1) implies that of (3).

(ii) Letting $U^0 = kG$, define for each $l \geq 0$,

$$e_l = \dim_k A_l/A_{l+1}, \ \ d_l = \dim_k U^l/U^{l+1} \, ,$$

so that

$$e_0, e_1, \dots, \ \ \ d_0, d_1, \dots$$

are two sequences of non-negative integers, both eventually zero since A and U are finite-dimensional over k. Note that

$$e_0 = 0, \ d_0 = 1, \ d_1 = d \, , \tag{4}$$

using (2), (13.14), (13.16), respectively. The exactness of (3) implies that for all $l \geq 0$,

$$e_l = d_{l+1} = dd_l \, , \tag{5}$$

since B has kG-rank d. Again using (2),

$$(AU^l) \, \alpha \subseteq BU^{l+1} \, ,$$

so that

$$AU^l \subseteq A_{l+1} \, ,$$

whence, by comparing codimensions,

$$r(d_0 + d_1 + \dots + d_{l-1}) \geq e_0 + e_1 + \dots + e_l \, . \tag{6}$$

(iii) Define two polynomials

$$g(t) = \sum_{l \geq 0} e_l t^l, \ f(t) = \sum_{l \geq 0} d_l t^l \, .$$

Now (5), together with the fact that $d_0 = 0$, implies that

$$tg(t) + (g(t) - 1) = df(t) \tag{7}$$

for all real t, while by (6),

$$\frac{rtf(t)}{1-t} \geq \frac{g(t)}{1-t} \tag{8}$$

for $0 < t < 1$. Eliminating $g(t)$ from (7) and (8), we have

$$(rt^2 - dt + 1)f(t) \geq 1 .$$

for $0 < t < 1$, and since $f(t)$ is positive in this range,

$$rt^2 - dt + 1 > 0, \quad \text{when } 0 < t < 1 . \tag{9}$$

Now the minimum of the quadratic $rt^2 - dt + 1$ occurs at $t = d/2r$, and since G is finite and non-trivial,

$$2r > r \geq d \geq 1,$$

using the exactness of (1). Thus, the minimum value of this quadratic occurs in the interval $(0,1)$, where it takes only positive values, by (9). It is thus positive everywhere and so has no real root. Hence the discriminant $d^2 - 4r$ must be negative, which proves the theorem.

2.   An example

This example (due to A.I. Kostrikin) is designed to give some idea of the accuracy of the bound in Theorem 1. For a fixed prime p and positive integer n, define

$$K_n = \langle X, Y \mid R, S, T \rangle ,$$

where

$$X = \{x_1, \ldots, x_n\}, \quad Y = \{y_1, \ldots, y_n\} ,$$

$$R = \{x_i^p = y_i^p = e \mid 1 \leq i \leq n\} ,$$

$$S = \{[x_i, y_j] = e \mid 1 \leq i, j \leq n\} ,$$

$$T = \{[x_i, x_j] = [y_i, y_j] \mid 1 \leq i < j \leq n\} .$$

Now $(K_n)_{ab}$ is clearly just the direct product of $2n$ copies of Z_p, whence $d(K_n) = 2n$. The number of relations is given by

$$r = 2n + n^2 + \frac{1}{2}n(n-1) = \frac{3}{2}(n^2 + n)$$

$$= \frac{3}{8}d^2 + \frac{3}{4}d .$$

This shows that the bound $r > d^2/4$ is reasonably accurate, provided we can prove that the K_n are all finite p-groups. It follows from the relations S and T that any (left-normed) commutator of length 3 in the generators must be the identity. Thus, K_n has class at most 2, that is, $K'_n \le Z(K_n)$. It then follows easily that K'_n is generated by commutators $[x,y]$, $x,y \in X$, and that each of these has order dividing p. As K'_n is abelian, it follows that K_n has order dividing $p^{(n^2 + 3n)/2}$.

3.   Related results

Note first of all that, while the main interest of the theorem is for finite p-groups, it actually holds for a broader class. Since the proof depends only on the existence of the sequence (1) with property (2), it is valid for any finite group for which Proposition 13.5 holds. An examination of §13.3 shows that the latter is valid for any presentation $G = <X \mid R>$ of a finite group with $|R| = r$ and $|X| = d$, where $p^d = |G : G'G^p|$.

Now the version given here is already an improvement on the original result (which had $r > (d-1)^2/4$). We mention two related results which hold under the same hypotheses as the theorem. The first of these is a genuine strengthening, and involves an invariant e of G defined as follows. Let $G_1 = G'G^p$, so that $|G : G_1| = p^d$, and let $G_2 = [G_1,G] G_1^p$ or $[G_1,G]G^p$ according as p is 2 or not. Then put $|G_1 : G_2| = p^e$.

Proposition 1. $r \ge d^2/2 + (-1)^p d/2 - e$, and

$$r > d^2/2 + (-1)^p d/2 - e - (e - (-1)^p d/2 - d^2/4) \, d/2 .$$

Proposition 2. $r > 2d - 1 - d/p$.

Remark. Using an elegant application of Proposition 1 to computational results of Havas, Richardson and Sterling, M.F. Newman has recently proved that $F(2,9)$, the last of the Fibonacci groups, is infinite.

Exercises

1. Prove that $|K_n| = p^{(n^2 + 3n)/2}$.

2. Investigate the class of groups defined in the first paragraph of §3.

3. Deduce the theorem from Proposition 1.

4. Show that Proposition 2 is an improvement of the theorem when $d \leq 7$ for large values of p.

Guide to the literature and references

Comments are given below on the material in each chapter in turn, and these embrace sources, alternative approaches and suggestions for further reading. The seven books referred to by authors' initials are of general interest. The lists of references are fairly minimal; a comprehensive bibliography is to be found in [LS].

[CM]   H.S.M. Coxeter and W.O.J. Moser, *Generators and relations for discrete groups*, 4th edition, Springer-Verlag, Berlin-Heidelberg-New York, 1979.

[J1]   D.L. Johnson, *Presentations of groups*, Cambridge University Press, 1976.

[J2]   D.L. Johnson, Topics in the theory of group presentations, Cambridge University Press, 1980.

[LS]   R.C. Lyndon and P.E. Schupp, *Combinatorial group theory*, Springer-Verlag, Berlin-Heidelberg-New York, 1977.

[M]   I.D. Macdonald, *The theory of groups*, Oxford University Press, 1968.

[MKS]   W. Magnus, A. Karrass and D. Solitar, *Combinatorial group theory*, Interscience, New York, 1966.

[R]   J.J. Rotman, *The theory of groups: an introduction*, 2nd edition, Allyn and Bacon, Boston, 1973.

1. The best introduction to group presentations is contained in Chapter 8 of [M]. The definition of free groups (via the adjoint of the forgetful functor) in Chapter 11 of [R] closely parallels our own, while interesting alternatives are given in Chapter 1 of [MKS] and Chapter 3 of [1]. Permutation groups are used to give a slick proof of the associative law for free products in [2], and the same method is used in Chapter 4 of [LS] to prove the normal form theorems for free products with amalgamation and HNN-extensions.

[1] J.R. Stallings, Group theory and three-dimensional manifolds, Yale University Press, New Haven 1971.

[2] B.L. van der Waerden, Free products of groups, Amer. J. Math. **70** (1948), 527 - 528.

2. Schreier's proof of Theoerm 1 appeared in [7], and the method is extended in [9] to prove the Kurosh theorem [4] for subgroups of free products. Both these theorems, as well as the (even harder) Grushko-Neumann theorem [2,5] now boast a number of

essentially topological proofs [R] (see also [6]), [1], [3], [8].

[1]   D.E. Cohen, *Combinatorial group theory: a topological approach,* Cambridge University Press, 1989.

[2]   I.A. Grushko, Über die Basen eines freien Produktes von Gruppen, *Mat. Sbornik,* N.S. **8** (1940), 169 - 182.

[3]   P.J. Higgins, *Notes on catergories and groupoids,* van Nostrand-Reinhold, New York, 1971.

[4]   A.G. Kurosh, Die Untergruppen der freien Produkte von beliebigen Gruppen. *Math. Ann.* **109** (1934), 647 - 660.

[5]   B.H. Neumann, On the number of generators of a free product, *J. London Math. Soc.* **18** (1943), 12 - 20.

[6]   J.J. Rotman, Covering complexes with applications to algebra, *Rocky Mountain J. of Math.* **3** (1973), 641 - 674.

[7]   O. Schreier, Die Untergruppen der freien Gruppen, *Abh. Math. Sem. Univ. Hamburg* **5** (1927), 161 - 183.

[8]   J.R. Stallings, A topological proof of Grushko's theorem on free products, *Math. Zeit.* **90** (1965), 1 - 8.

[9]   A.J. Weir, The Reidemeister-Schreier and Kurosh subgroup theorems, *Mathematica* **3** (1956), 47 - 55.

3. The Nielsen-Schreier theorem is proved for finitely-generated subgroups in [2] (translated by J.C. Stillwell in [3]), and the method is extended to arbitrary subgroups in [1]. The proof given here is taken from Chapter 1 of [LS] as are some of the corollaries. Other corollaries and Example 2 are from Chapters 2 and 3 of [MKS] and Chapter 5 of [4].

1.   H. Fedorov and B. Jónsson, Some properties of free groups, Trans. Amer. Math. Soc. **68** (1950), 1 - 27.

2.   J. Nielsen, Om Regning med ikke kommutative Faktoren og deus Anvendelse i Gruppeteorien, Mat. Tidssk. B (1921), 77 - 94.

3.   J. Nielsen, Collected Mathematical papers, Birkhäuser 1986.

4.   E.V. Schenkman, Group theory, van Nostrand, Princeton 1965.

4. Tietze transformations first appeared in [3] and van Kampen diagrams in [1]. The latter form the starting point of small cancellation theory, which was pioneered in [2] and

has since become one of most important methods of combinatorial group theory. See Chapter 7 of [J2] for an introduction and Chapter 5 of [LS] for a thorough treatment.

[1]   E.R. van Kampen, On some lemmas in the theory of groups, Amer. J. Math. **55** (1933), 268 - 273.

[2]   R.C. Lyndon, on Dehn's algorithm, Math. Ann. **166** (1966), 208 - 228.

[3]   H. Tietze, Über die topologischen Invarianten mehrdimensionaler Mannigfaltig-keiten, Monatsh. Math. Phys. **19** (1908), 1 - 118.

5. Presentations for the dihedral and quaternionic (or dicyclic) groups are given in Chapter 7 of [1], whic also contains an alternative presentation for S_n and a related one for A_n. Groups of symmetries of geometrical configurtions are studied in detail in [CM] and [3], and polynomials under substitution in [2].

[1]   R.D. Carmichael, An introduction to the theory of groups of finite order, Dover, New York 1956.

[2]   D.L. Johnson, The group of formal power series under substitution, J. Austral. Math. Soc. (series A) **45** (1988), 296 - 302.

[3]   R.C. Lyndon, Groups, and geometry, LMS Lecture Notes no. 101, Cambridge University Press, 1985.

6. Another proof of the basis theorem appears in [M] and [R], while that given here (including the Invariant Factor theorem for matrices) appears in a more general form in §16 of [1]. Corollary 2 is Theorem 8.16 of [M], where a direct proof is given.

[1]   C.W. Curtis and I. Reiner, Representation theory of finite groups and associative algebras, Interscience, New York 1962.

7. The definition and basic properties of the multiplicator are to be found in [14] (see [19] for a readable survey) and some finite groups with trivial multiplicator and non-zero deficiency appear in [15]. The deficiency problem for metacyclic groups is solved in [16] and [1], and the spectral sequence argument in the latter is obviated by the use of central stem extensions in [2]. The groups of Mennicke, Macdonald and Wamsley appear in [12], [11] and [17], respectively, and the first of these has been embedded in a much larger (5-ply infinite) class by Post [13]. The $J(a,b,c)$ are studied in [8], which is based on [18]. $F(2,8)$ and $F(2,10)$ are proved to be infinite in [3], and the bound for $|F(2,9)|$ is given in [5]. The Fibonacci groups are introduced in [9] and studied further in [6], [7] and several papers by Campbell and Robertson (see the references in [10]). These authors

have also found two-generator, two-relation presentations for the $SL(2,p)$ and $SL(2,8)$ [4], and the latter is the only known example of an interesting simple group. A more comprehensive list of references is given in the survey article [10].

[1] F.R. Beyl, The Schur multiplicator of metacyclic groups, *Proc. Amer. Math. Soc.* **40** (1973), 413 - 318.

[2] F.R. Beyl and M.R. Jones, Addendum to 'the Schur multiplicator of metacyclic groups', *Proc. Amer. Math. Soc.* **43** (1974), 251 - 252.

[3] A.M. Brunner, The determination of Fibonacci groups, *Bull. Austral. Math. Soc.* **11** (1974), 11 - 14.

[4] C.M. Campbell and E.F. Robertson, Two-generator two-relation presentations for special linear groups, *Bull. London Math. Soc.* **12** (1980), 17 - 20.

[5] G. Havas, J.S. Richardson and L.S. Sterling, The last of the Fibonacci groups, *Proc. Royal Soc. Edinburgh* **83A** (1979), 199 - 203.

[6] D.L. Johnson, Extensions of Fibonacci groups, *Bull. London Math. Soc.* **7** (1974), 101 - 104.

[7] D.L. Johnson, Some infinite Fibonacci groups, *Proc. Edinburgh Math. Soc.* **19** (1975), 311 - 314.

[8] D.L. Johnson, A new class of 3-generator finite groups of deficiency zero, *J. London Math. Soc.* **19** (1979), 59 - 61.

[9] D.L. Johnson, J.W. Wamsley and D. Wright, The Fibonacci groups, *Proc. London Math. Soc.* **29** (1974), 577 - 592.

[10] D.L. Johnson and E.F. Robertson, Finite groups of deficiency zero, in *Homological group theory* (ed. C.T.C. Wall), Cambridge Unviersity Press, 1979.

[11] I.D. Macdonald, On a class of finitely-presented groups, *Canad. J. Math.* **14** (1962), 602 - 613.

[12] J. Mennicke, Einige endliche Gruppen mit drei Erzeugenden und drei Relationen, *Arch. Math.* **10** (1959), 409 - 418.

[13] M.J. Post, Finite three-generator groups with zero deficiency, *Comm. Alg.* **6** (1978), 1289 - 1296.

[14] J. Schur, Untersuchungen über die Darstellung der endlichen Gruppen durch gebrochene lineare Substitutionen, *J. Reine Angew. Math.* **132** (1907), 85 - 137.

[15] R.G. Swan, Minimal resolutions for finite groups, *Topology* **4** (1965), 193 - 208.

[16] J.W. Wamsley, The deficiency of metacyclic groups, *Proc. Amer. Math. Soc.* **24** (1970), 724 - 726.

[17] J.W. Wamsley, A class of three-generator three-relation finite gorups, *Canad. J. Math.* **22** (1970), 36 - 40.

[18] J.W. Wamsley, Some finite groups with zero deficiency, *J. Austral. Math. Soc.* **18** (1974), 73 - 75.

[19] J. Wiegold, The Schuer multiplier: an elementary approach, *in* Groups-St. Andrews 1981, LMS lecture notes no. 71, Cambridge University Press 1982, pp 137 - 154.

8. The coset enumeration process made its debut in [3], and a description of a modified method yielding a presentation of the subgroup is given in Chapter 2 of [CM]. For more recent developments, see the surveys [2] and [1].

[1] J. Leech, Coset enumeration, *in* Computational Group theory, Academic Press, London 1984, pp 3 - 18.

[2] J. Neubüser, An elementary introduction to coset table methods in computational group theory, in Groups-St. Andrews 1981, LMS lecture notes no. 71, Cambridge University Press 1982, pp. 1 - 45.

[3] J.A. Todd and H.S.M. Coxeter, A practical method for enumerating cosets in an abstract finite group, Proc. Edinburgh Math. Soc. (2) **5** (1936), 25 - 36.

9. Braid groups are treated in [2], [3] and [5]. The triangle groups were first studied intensively in [4] and a more up to date treatment is to be found in [CM]. A purely algebraic proof that $D(l,m,n)$ is infinite if and only if $1/l + 1/m + 1/n \leq 1$ is given in [6], and a slick graph-theoretical proof appears in [8], along with a generalization. Free products and free products with amalgamation form the topic of Chapter 4 of [MKS] and, together with HNN-extensions, of Chapter 4 of [LS]. For the Schur multiplicator, see the references for Chapter 7. The Baumslag-Solitar group appeared in [1], and the result of Exercise 16 is to be found in [7].

[1] G. Baumslag and D. Solitar, Some two-generator one-relator non-Hopfian groups, Bull. Amer. Math. Soc. **68** (1962), 199 - 201.

[2] J.S. Birman, Braids, links and mapping class groups, Ann. of Math. Studies no. 82, Princeton University Press, Princeton 1974.

[3]   G. Burde and H. Zieschang, Knots, de Gruyter, Berlin 1985.

[4]   R. Fricke and F. Klein, Vorlesungen über die Theorie der automorphen Funktsionen, Vol. 1, Teubner, Leipzig 1897.

[5]   W. Magnus, Braid groups: a survey, in Lecture Notes in Math. no. 372, Springer-Verlag, Berlin 1974, pp. 463 - 487.

[6]   G.A. Miller, Groups defined by the orders of two generators and the order of their product, Amer. J. Math. **24** (1902), 96 - 100.

[7]   C. Squier, On certain 3-generator Artin groups, Trans. Amer. Math. Soc. **302**:1 (1987), 117 - 124.

[8]   R.M. Thomas, Cayley graphs and group presentations, Math. Proc. Camb. Phil. Soc. **103**:3 (1988), 385 - 387.

10. The classical treatment of group extensions (using cohomology) is [1]. For group theory with a homological flavour, see the notes [3] and the collection [6]. Proposition 2 is due to J.A. Green [2]. For a more accurate estimate than Proposition 3 for the number of p-groups, see [4] and [5].

[1]   S. Eilenberg and S. Maclane, Comology theory in abstract groups I, Ann. of Math. **48** (1947), 326 - 341.

[2]   J.A. Green, On the number of automorphisms of a finite group, Proc. Roy. Soc. London, Ser. A **237** (1956), 574 - 581.

[3]   K.W. Gruenberg, Cohomological topics in group theory, Lecture Notes in Math. no. 143, Springer-Verlag, Berlin 1970.

[4]   G. Higman, Enumerating p-groups I: inequalities, Proc. London Math. Soc. (3) **10** (1960), 24 - 30.

[5]   C.C. Sims, Enumerating p-groups, Proc. London Math. Soc. (3) **15** (1965), 151 - 166.

[6]   C.T.C. Wall (ed.), Homological group theory, LMS Lecture Notes no. 136, Cambridge University Press 1979.

11. The free differential calculus was introduced in [2], which also contains the proofs given here of Proposition 5 and 6 (due originally to Blanchfield and Lyndon, respectively: see Chapter 3 of [1]). Lemma 4 is Proposition 3.1 of [3].

[1]   J.S. Birman, Braids, links and mapping class groups, Ann. of Math. Studies no. 82, Princeton University Press, Princeton 1974.

[2]   R.H. Fox, Free differential calculus I, Ann. of Math. **57** (1953), 547 - 560.

[3]   K.W. Gruenberg, Cohomological topics in group theory, Lecture Notes in Math. no. 143, Springer-Verlag, Berlin 1970.

12. The algorithm is described in [1], which contains further examples. Chapter 3 of [MKS] contains an elegant exploitation of Nielsen transformations that computes the invariant factors of G/G' and of G'/G'' at the same time.

[1]   J.R. Howse and D.L. Johnson, An algorithm for the second derived factor, *in* Groups-St. Andrews 1981, LMS lecture notes no. 71, Cambridge University Press 1982, pp. 237 - 243.

13. The elementary theory of p-groups is covered in [M] and {R}. Proposition 2 is proved in [1].

[1]   J.W. Wamsley, Computation in nilpotent groups (theory), *in* Proc. Second Internat. Conf. Theory of Groups (Canberra 1973), Lecture Notes in Math. no. 372, Springer-Verlag, Berlin 1974, pp. 691 - 700.

14. The nilopotent quotient algorithm was conceived by I.D. Macdonald [2]. The description given here is due to M.F. Neumann [3]. See [1] for a survey of its applications, and [4] for the improvement.

[1]   G. Havas and M.F. Newman, Application of computers to questions like those of Burnside, *in* Lecutre Notes in Math. no. 806, Springer-Verlag, Berlin 1980, pp. 211 - 230.

[2]   I.D. Macdonald, A computer application to finite p-groups, J. Austral. Math. Soc. **17** (1974), 102 - 112.

[3]   M.F. Newman, Some group presentations and enforcing the associative law, *in* Lecture Notes in Comput. Sci. 229, Springer-Verlag, Berlin 1986, pp. 228 - 237.

[4]   M.R. Vaughan-Lee, An aspect of the nilpotent quotient algorithm, *in* Computational Group Theory, Academic Press, London 1984, pp. 75 - 83.

15. The theorem first appeared in [1] and this proof in [5]. The example is in [3]. Propositions 1 and 2 are to be found in [2] and [4] respectively.

[1] E.S. Golod and I.R. Shafarevich, On the class field tower, Izv. Akad. Nauk SSSR **28** (1964), 261 - 272.

[2] W. Gaschütz and M.F. Newman, On presentations of finite p-groups, J. Reine Angew. Math. **245** (1970), 172 - 176.

[3] B. Huppert, Endliche Gruppen I, Springer-Verlag, Berlin 1967.

[4] D.L. Johnson, A property of finite p-groups with trivial multiplicator, Amer. J. Math. **98**: 1 (1976), 105 - 108.

[5] P. Roquette, On class field towers, *in* Algebraic Number Theory, Academic Press, London 1967, pp. 231 - 249.

INDEX

Dramatis Personae
(in approximate order of appearance)

$\mathbb{N}$	natural numbers	$\mapsto$	effect of mapping on an
$\mathbb{Z}$	integers		element
$\mathbb{Q}$	rationals	inc	inclusion mapping
$\mathbb{R}$	reals	nat	natural homomorphism
$\mathbb{H}$	quaternions	$\cong$	isomorphism
$\aleph_0$	aleph-null $:= \mid \mathbb{N} \mid$	1_x	identity mapping on set x
$r(F)$	rank of free group F	Ker θ	kernel of mapping θ
$\mathbb{Z}_n$	ring of integers mod n	Im θ	image of mapping θ
$GF(p)$	$\mathbb{Z}_p$, p a prime		
$\mathbb{Z}G$	group-ring	H	first discrete Heisenberg group
kG	group-algebra	$Q(+)$	additive group of rationals
		Aut A	automorphism group of A
Map (X,G)	mappings from X to G	Im A	group of inner automorphisms of A
Hom (F,G)	homomorphism from F to G	Z_n	cyclic group of order $n \in \mathbb{N}$
$<X>$	subgroup generated by X	Z	infinite cyclic group
$\overline{R}$	normal closure of R	$F(X)$	free group on set X
$<X \mid R>$	group presentation	Sym X	symmetric group on X
X^-	$\{x^{-1} \mid x \in X\}$	S_n	symmetric group
$X^{\pm}$	$X \cup X^-$	A_n	alternating group
$\mid X \mid$	cardinality of set X	$G_n(p)$	power-series group
\	set-theoretic difference		
$\cap$	intersection	$D(l,m,n)$	von Dyck group
$\cup$	union	$\Delta(l,m,n)$	triangle group
$\overset{\cup}{\cdot}$	disjoint union	B_n	braid group
$\varnothing$	empty set	$\hat{B}_n$	circular briad group
		Q_{2_n}	generalized quaternion group
$\rightarrow$	mapping	D_{2_n}	dihedral group
$>$	mapping whose existence is	$F(+,n)$	Fibonacci group
	alleged	$GL(n,R)$	general linear group
$\twoheadrightarrow$	surjection	$M(a,b,c)$	Mennicke groups
$>\!\!\rightarrow$	injection	$W_{\pm}(a,b,c)$	Wamsley groups

Mac (a,b)	Macdonald groups	$N(H)$	normalizer of subgroup H
$J(a,b,c)$	my groups	$C(a)$	centralizer of element a
K_n	Kostrikin group	$[X,Y]$	(subgroup generated by) set of
$G_n(w)$	cyclically-presented group		commutators $\{x^{-1}y^{-1}xy \mid x \in X, y \in Y$
$\pi_1(s)$	fundamental group of space S	G_{ab}	G/G'
		$M(G)$	Schur multiplicator of group G
$\oplus$	direct sum	def G	deficiency of G
$\times$	direct product	$d(G)$	minimal number of generators of G
]	semi-direct product	$H^n(G,A)$	cohomology group
$*$	free product		
$G*_K H$	free product with amalgamation	$l(w)$	length of word w
$G*_K \phi$	HNN-extension	$\mathcal{F}$	free groups
$\otimes$	Kronecker product	$\mathcal{A}$	abelian groups
$\leq$	less than or equal, is a	$\mathcal{N}$	nilpotent group
	subgroup of	$_R\chi$	redisually χ-groups
$<$	less than, is a proper subgroup of	δ_{xy}	Kronecker delta
$\triangleleft$	is a normal subgroup of	$\sum'$	finite sum
$A:=B$	A is defined to be B	I_n	identity matrix
		$\phi(n)$	Euler totient function
G^p	subgroup of G generated	$n!$	n factorial
	by pth powers	$\exists!$	there is one and only one (onne)
$\Phi(G)$	Frattini subgroup	Stab	stabilizer
G'	derived group (commutator	PCP	power-commutation presentation
	subgroup) of G	e	empty word, identity element
$Z(G)$	centre of group G	E	trivial group

LONDON MATHEMATICAL SOCIETY STUDENT TEXTS

Managing editor: Professor E.B. Davies, Department of Mathematics,
King's College, Strand, London WC2R 2LS